Nature, Environment and Poetry

The environmental challenges facing humanity in the twenty-first century are not only acute and grave; they are also unprecedented in kind, complexity and scope. Nonetheless, the political response to problems such as climate change, biodiversity loss and widespread pollution continues to fall short. To address these challenges it seems clear that we need new ways of thinking about the relationship between humans and nature, local and global, and past, present and future. One place to look for such new ideas is in poetry, designed to contain multiple levels of meaning at once, challenge the imagination, and evoke responses that are based on something more than scientific consensus and rationale.

This ecocritical book traces the environmental sensibilities of two Anglophone poets: Nobel Prize-winner Seamus Heaney (1939–2013) and British Poet Laureate Ted Hughes (1930–1998). Drawing on recent and multifarious developments in ecocritical theory, it examines how Hughes's and Heaney's respective poetics interact with late twentieth century developments in environmental thought, focusing in particular on ideas about ecology and environment in relation to religion, time, technology, colonialism, semiotics and globalisation.

This book is aimed at students of literature and environment, the relationship between poetry and environmental humanities, and the poetry of Ted Hughes or Seamus Heaney.

Susanna Lidström is a postdoctoral researcher at the Division of History of Science, Technology and Environment, with the Environmental Humanities Laboratory, at the KTH Royal Institute of Technology in Stockholm, Sweden.

Routledge Environmental Humanities

Series editors: Iain McCalman and Libby Robin

"After the theoretical upheavals of the past five years or so, what is now needed in ecocritical scholarship is a book that takes account of these developments *and* applies them to texts we thought we knew. This is exactly what Lidström does, in an admirably clear, compelling style."

–Adeline Johns-Putra, Chair, Association for the Study of Literature and Environment, UK and Ireland

"Lidström has written a lucid, wide-ranging ecocritical study of the most distinguished poets of the British Isles in the late 20th century. Her chief innovation is to orchestrate a subtle interplay between the poets' collections and the theoretical frameworks they seem to require, rather than bashing them all into a single ecopoetic mould. The result is wonderfully illuminating."

–Greg Garrard, University of British Columbia, Okanagan, USA

"Susanna Lidström creates a fascinating and timely study of Hughes and Heaney as ecopoets. Arguing that environmental issues have 'changed the mind of poetry', she draws exciting new insights from ecosemiotics, postcolonial ecocriticism, and theories of local and global to create an innovative and important new work."

–Yvonne Reddick, University of Central Lancashire, UK

"Lidström's clear and accessible study illuminates two versions of ecopoetics, one committed to anti-anthropocentric revelation of what lies beyond social and cultural constructions of nature, the other invested in the intertwining of nature and culture. Examining the poetry of Hughes and Heaney through six distinct methodological and thematic lenses, she valuably highlights the diversity of contemporary ecocritical practices."

–Lynn Keller, University of Wisconsin–Madison, USA

The *Routledge Environmental Humanities* series is an original and inspiring venture recognising that today's world agricultural and water crises, ocean pollution and resource depletion, global warming from greenhouse gases, urban sprawl, overpopulation, food insecurity and environmental justice are all *crises of culture*.

The reality of understanding and finding adaptive solutions to our present and future environmental challenges has shifted the epicentre of environmental studies away from an exclusively scientific and technological framework to one that depends on the human-focused disciplines and ideas of the humanities and allied social sciences.

We thus welcome book proposals from all humanities and social sciences disciplines for an inclusive and interdisciplinary series. We favour manuscripts aimed at an international readership and written in a lively and accessible style. The readership comprises scholars and students from the humanities and social sciences and thoughtful readers concerned about the human dimensions of environmental change.

Rethinking Invasion Ecologies from the Environmental Humanities
Jodi Frawley and Iain McCalman

The Broken Promise of Agricultural Progress
An environmental history
Cameron Muir

The Biosphere and the Bioregion
Essential writings of Peter Berg
Cheryll Glotfelty and Eve Quesnel

Sustainable Consumption and the Good Life
Interdisciplinary perspectives
Edited by Karen Lykke Syse and Martin Lee Mueller

The Anthropocene and the Global Environmental Crisis
Rethinking Modernity in a new Epoch
Edited by Clive Hamilton, Christophe Bonneuil and François Gemenne

Nature, Environment and Poetry
Ecocriticism and the poetics of Seamus Heaney and Ted Hughes
Susanna Lidström

Whole Earth Thinking and Planetary Coexistence
Ecological wisdom at the intersection of religion, ecology, and philosophy
Sam Mickey

Endangerment, Biodiversity and Culture
Edited by Fernando Vidal and Nélia Dias

Nature, Environment and Poetry

Ecocriticism and the poetics of Seamus Heaney and Ted Hughes

Susanna Lidström

Routledge
Taylor & Francis Group
LONDON AND NEW YORK

earthscan
from Routledge

First published 2015
by Routledge

2 Park Square, Milton Park, Abingdon, Oxon OX14 4RN
711 Third Avenue, New York, NY 10017, USA

First issued in paperback 2017

Routledge is an imprint of the Taylor & Francis Group, an informa business

British Library Cataloguing-in-Publication Data
A catalogue record for this book is available from the British Library

Library of Congress Cataloging-in-Publication Data
A catalog record for this book has been requested

ISBN: 978-1-138-77524-4 (hbk)
ISBN: 978-1-138-74358-8 (pbk)

Typeset in Goudy
by HWA Text and Data Management, London

Contents

Acknowledgements

I want to thank Libby Robin for her enthusiastic support of this book from start to finish. I also want to thank everyone else at the Division of History of Science, Technology and Environment at KTH Royal Institute of Technology in Stockholm, and in particular Sverker Sörlin and Nina Wormbs, for their perspectives and encouragement. My gratitude also goes to those who advised me while writing the dissertation that is the prequel to this book, including my supervisor at King's College London, Richard Kirkland, and my external supervisor Greg Garrard.

A special thank you to Joana Medeiros for letting me use her beautiful painting for the cover of the book.

Many thanks also to my family and friends for their support, and especially to Erik Olofsson for inquiring and engaging discussions that have made me think in new ways about the broader framework for the different topics of this book.

Introduction

In 2008, Seamus Heaney stated that 'environmental issues have to a large extent changed the mind of poetry' (O'Driscoll, 2008: 407). The remark was picked up by columnist Robert Butler in 2011 to make the point that the 'idea of going green' is about much more than changing our light bulbs:

> Every big scientific moment is also a cultural one. The Lisbon earthquake that killed an estimated 30,000 people in 1755 gave birth to the science of seismology. It also inspired writings by Kant, Rousseau and, most famously, Voltaire, who describes the earthquake in 'Candide', and the impact it had on the notion that there was a benevolent God watching over 'the best of all possible worlds'. A century later, the ideas in Darwin's 'Origin of Species' would be played out, absorbed and contested in the novels of George Eliot and Thomas Hardy. Fifty years after that, Einstein's Theory of Relativity paved the way for modernism in all the arts.
>
> (Butler, 2011)

Likewise, the growing realisation and acknowledgement of accelerating human impact on the natural environment in the twentieth century is now having a profound effect on how we think about the relationship between human societies and the natural world (see e.g. Steffen et al., 2007). As Butler suggests, this change can be compared to other moments in human history when our worldview was revolutionised in response to scientific discoveries, such as the Renaissance or Reformation or, as Schellnhuber has pointed out elsewhere, the Copernican revolution (Schellnhuber, 1999).

As Heaney recognises, this rethinking of human–earth relations is also changing the stories we tell about the world, including those in poetry. But how do poems in particular reflect and contribute to this major change in our environmental imaginations? How does the context of growing concern for the environment affect the way in which we read poems, especially ones pertaining, more or less directly, to the relationship between human and non-human nature?

This book explores these two questions by analysing ideas about nature and environment in the poetry by Seamus Heaney and Ted Hughes, two important poets in the Western, Anglophone world whose respective careers have developed parallel to the rise of modern environmentalism in the second half of the twentieth century. Both have been referred to as 'ecopoets', and are well-known for poems about place, animals, and landscape. Both also enjoy growing popularity, especially in Ireland and the UK. As our ideas about nature and environment are increasingly scrutinised, this book aims to unpack the idea of the 'ecopoet', and ask specifically what *kind* of ecopoets Hughes and Heaney may be, and also point to the question of what kinds of ecopoets there are in general. If poems by Hughes are taught as part of a class on poetry and environment, what does that say about our ideas regarding what the term 'environment' actually denotes? Or if Heaney and Hughes are listed alongside each other as examples of so-called ecopoets, does that mean that they share some basic understanding of the relationship between humans, poetry, and the natural world? Or is 'ecopoet' a label that can house disparate views of the relationship between these concepts and categories?

In its aim to investigate these questions, this book is structured around a series of independent case studies rather than as a continuous and comparative analysis of Heaney's and Hughes's poetics following a specific temporal or thematic progression or orientation. Each chapter (or case study) pairs a particular ecocritical theory with a selected set or collection of poems by either poet. This structure makes it possible to identify several important individual themes and ideas in Hughes's and Heaney's respective poetry and examine them in their own right. The conclusion at the end of the book then draws on the results from the different chapters to discuss the idea of an 'ecopoetics' from a broader point of view. The structure of the book is thus more circular than linear; the central question, regarding what kind of ideas about nature and environment inform Hughes's and Heaney's respective poetry, remains the key issue, but is approached from a variety of directions and theoretical perspectives.

The field of ecocriticism, defined as the study of literature and the environment, developed in response to growing recognition and awareness of environmental crises in many parts of the world in the decades after the Second World War, beginning with a focus on American and English literary traditions. Initially inspired by nature writers such as Henry David Thoreau and environmentalists like Rachel Carson, and academically by scholars such as Leo Marx and Raymond Williams, ecocriticism has developed over the past few decades into a diverse field of enquiry, geographically as well as thematically. In a narrow sense, ecocritical analyses examine how narratives and other forms of cultural representation influence and are affected by environmental concerns and crises. In a wider sense, ecocriticism studies the relationship between nature and culture from a much broader perspective, beyond the idea of an ongoing crisis.

The first explicit attempts to introduce ecocriticism, or green literary criticism, as a new academic field were made in the 1990s. In the UK, *Romantic Ecology: Wordsworth and the Environmental Tradition* by Jonathan Bate was published in 1991. One year later, across the Atlantic, the Association for the Study of Literature and the Environment (ASLE) was formed, and in 1993 the first issue of their journal, *Interdisciplinary Studies in Literature and Environment*, or *ISLE*, came out. ASLE remains the largest professional association for ecocritical scholars, now with national and regional affiliations across the globe, including several branches with their own journals, such as ASLE-UKI's (UK and Ireland) *Green Letters*, and EASLCE's (European Association for the Study of Literature, Culture and Environment) *Ecozon@*. In 1995 Lawrence Buell's influential *The Environmental Imagination: Thoreau, Nature Writing, and the Formation of American Culture* was published, often referred to as a founding text of ecocriticism.

While a focus on America and the UK seems to suggest a linear, progressive development of environmental literary criticism (often referred to as the 'wave' model; for a recent critical account see Bergthaller et al., 2014), originating and at least initially confined to the Western, Anglophone world, looking beyond the specific term ecocriticism a different picture emerges. Buell has also recognised that, while in some respects ecocriticism may be considered a new field, it builds on much older traditions in the sense that the 'idea of nature' has always been more or less prominent in literary studies (2005: 2). The field is also broader and more diverse than the focus on America and the UK would suggest. A recent overview suggests that, if 'we abandon received ideas about the field's development, particularly the dominant historiographical narrative of ecocriticism that has emerged from the United States', a new history of environmental literary studies would likely emerge to tell a different story about the background and increasing significance of ideas we now consider 'ecocritical' (Bergthaller et al., 2014). Rob Nixon begins to introduce such an alternative view and history of ecocriticism in his study from 2011, *Slow Violence and the Environmentalism of the Poor*.

While green literary criticism may be much more diverse than is often appreciated from within the field as it has developed and been defined so far, there is also significant multiplicity already within that field. Since the mid-1990s, a growing number of theoretical perspectives and combinations have developed as subfields to Anglo-American ecocriticism, to bring out specific environmental themes in different kinds of texts. This book focuses on six such subfields, namely *environment and religion*; *postmodern nature*; *technology and landscape*; *postcolonial ecocriticism*; *ecosemiotics*; and *globalisation and sense of place*. This compilation of theories is meant to reflect Buell's description of literary ecotheory 'unfolding less as a story of dogged recalcitrance [...] than as a quest *for* adequate models of inquiry from the plethora of possible alternatives that offer themselves from whatever disciplinary quarter', suggesting that environmental literary studies are better understood 'less as a monolith than as a concourse of discrepant practices' (2005: 10).

Among these divergent ideas and approaches, the focus on two poets, maintained throughout the book in alternating chapters, is intended to provide a sense of continuity. The different theoretical approaches are chosen for their potential to address and bring out specific and important themes in each poet's work. The diverse perspectives are tied together in the end in a concluding discussion that considers the results of the different chapters from a wider perspective and in relation to each other. While drawing on the preceding analyses of Heaney's and Hughes's respective ecopoetics, the conclusion addresses the notion of ecopoetry from a broader perspective, and suggests a more nuanced conceptualisation of the relationship between nature, environment and poetry than the label 'ecopoetry' might suggest.

Nature and religion

The first scholar to suggest that there was a causal relationship between religion and environmental crisis was Lynn White, in his pioneering article 'The Historical Roots of our Ecologic Crisis', published in *Science* in 1967. White argued that the North American and Western European idea that scientific knowledge equals technological power over nature, an equation that merges theoretical and empirical approaches to the environment, is as recent as from around 1850. Before then, White stated, science was 'aristocratic, speculative, intellectual in intent; technology was lower-class, empirical, action-oriented' (1967: 1204). The sudden fusion of these traditions, and the rapid acceptance of their direct relationship as the ordering principle of the relationship between humans and their environment, White suggested, 'may mark the greatest event in human history since the invention of agriculture, and perhaps in nonhuman terrestrial history as well' (1967: 1203). A consideration of the brief history of this idea highlights, according to White, the shallowness and inadequacy of the dominant responses to environmental crisis so far, as they address merely fractions or symptoms of much more fundamental concerns. Rather than inventing technological solutions to narrowly defined environmental problems, White argued that instead we need to address the underlying patterns of thought that are the cause of these recent changes.

To do so, White turns to the question of religion, and in particular to Christianity, which he suggests is the belief-system that has enabled and promoted the merging of science and technology, and which therefore may be considered the underlying cause of contemporary environmental crises:

> The victory of Christianity over paganism was the greatest psychic revolution in the history of our culture. It has become fashionable today to say that, for better or worse, we live in 'the post-Christian age.' Certainly the forms of our thinking and language have largely ceased to be Christian, but to my eye the substance often remains amazingly akin to that of the past. Our daily habits of action, for example, are dominated by an implicit faith in perpetual progress which was unknown either to Greco-Roman antiquity or to the Orient. It is rooted in, and is indefensible apart from, Judeo-Christian theology. The fact that Communists share it merely helps to show what can be demonstrated on many other grounds: that Marxism, like Islam, is a Judeo-Christian heresy. We continue today to live, as we have lived for about 1700 years, very largely in a context of Christian axioms.
>
> (1967: 1205)

The first chapter in this book takes White's article as a starting point for an analysis of Hughes's poetry collection *Crow: From the Life and Songs of the Crow.*

The motivation behind this pairing is both historiographical and theoretical: *Crow* was published in 1970, only three years after White's article, and the two publications, though through very different forms, express a remarkably similar set of ideas. The *Crow* poems revolve around the same critical points as White's argument, namely the relationship between Christianity, science and technology, and environmental degradation.

While this chapter suggests that *Crow* was to a certain extent a reflection of particular environmental ideas that were emerging around this time, in different forms and shapes and on different sides of the Atlantic, it also shows that while White and Hughes begin from common ground, they proceed in different directions. While White suggests a revision of Christianity as a belief-system that will make it more responsible towards environmental change, Hughes, through the figure of Crow, wholly and emphatically renounces Christianity, and instead suggests a radical alternative. This alternative reflects and relates to ideas developing elsewhere in Western societies at the same time, and can be seen as part of an emerging trend towards different forms of nature spirituality as a way to reformulate human–nature relations, as outlined by Bron Taylor in *Dark Green Religion: Nature Spirituality and the Planetary Future* (2010). Taylor's study helps to show how *Crow* forms part of a particular reconsideration and revision of the relationship between nature, religion and environmentalism towards a more spiritual dimension, present in different forms in popular culture since at least the 1960s. Hughes's *Crow* collection both reflects and informs, enacts and conceptualises this revision.

Postmodern ecocriticism

While American ecocriticism emerged at least in part as a reaction against post-structuralist and postmodern literary studies, attempts have since been made to combine postmodern theory and environmental literary criticism to develop something like a postmodern ecocritical practice. One proponent of this development is Serpil Oppermann, who argues that while early ecocriticism 'signalled a new and a promising hermeneutical horizon in our interpretations and understanding of the natural world and literature', its potential is limited as long as it confines itself to 'the theoretically discredited parameters of literary realism' (2006: 103). To challenge this confinement Oppermann wants to conflate ecocritical and postmodern theory, in order 'first to critique ecocriticism's realist orientation as being inappropriate for literary theorizing, second to provide a valid account of postmodernism which is more reconstructive than deconstructive in its ecological field of vision' (2006: 104).

Oppermann argues that the attempt to align early ecocriticism with a realist epistemology in order to turn it into 'an open field of inquiry' is paradoxical because it discredits the relationship between interpretation and language (2006: 103). The opposition between some ecocritics and certain postmodern and deconstructive literary theories that Oppermann addresses is summarised by Timothy Clark in *The Cambridge Introduction to Literature and Environment*:

> A persistent target of environmental critics, especially in the 1990s, was a view of language which they attributed loosely to 'postmodern', 'deconstructive' or 'post-structuralist' theory. This is the claim that language forms a kind of cultural prison, confining its users to the specific conceptions and presumptions it projects – an argument encountered often in third- or fourth-hand accounts of thinkers such as Jacques Derrida or Jean-François Lyotard, even though their actual arguments are very different. Ecocritics saw themselves as resisting claims that 'no authoritative and definitive expression or conception of reality is possible' and that 'all we can ever perceive about the world are shadows, and that we can never escape our particular biases'. At times this led to what are now acknowledged as ecocritical caricatures of so-called 'post-structuralism'. The relatively uncontroversial argument that human beings cannot know reality absolutely, without some cultural presuppositions, was sometimes taken to be the patently silly one of denying the existence of reality altogether.
>
> (2011: 46)

Oppermann's attempt to bring ecocritical and postmodern theory into conversation with each other have since been followed up by other critics, such as Timothy Morton in *Ecology without Nature* (2007) and Cary Wolfe in *What is Posthumanism?* (2009). Morton states that his book 'is inspired by the way in which deconstruction searches out, with ruthless and brilliant intensity, points of contradiction and deep hesitation in systems of meaning' (2007: 6), a definition elaborated on in the second chapter of this book.

Chapter 2 examines how Oppermann's outline of a postmodern ecocritical methodology may be adapted for poetry analysis through a reading of Heaney's famous 'bog poems', most of which are in the collection *North* (1975). Written during the height of the Troubles in Northern Ireland in the 1970s, when Heaney experienced considerable pressure, from himself as well as others, to provide some kind of commentary on the events taking place in his home region, *North* is an unexpected reply. Rather than commenting directly on the contemporary situation in Northern Ireland, the bog poems are centred on descriptions of bodies from ancient history, unearthed on one of the islands of Denmark, that seem to have been victims of rituals performed in the name of land and religion. Conserved for centuries in peat moors, around the time that Heaney wrote *North* several such bodies came to the surface in both Denmark and Ireland, and were subjected to examination and ensuing theories about their different destinies. Heaney's bog poems take this long temporal and geographical detour, before approaching the contemporary violence in Northern Ireland.

North is a careful and self-conscious balancing act between artistic obligation and integrity. When asked about the former, Heaney once replied that 'I think the poet who didn't feel the pressure at a politically difficult time would be either stupid or insensitive', but he also acknowledges 'the crucial matter of artistic ability' – whether poets are able to take on a politically sensitive subject matter without abandoning the integrity of their art (Cole, 1997). For Heaney, the bog poems were a way for him to respond to (rather than address, a distinction he makes in the same interview) the Troubles, without feeling that he compromised his artistic integrity or vision.

Reading the bog poems from the perspective of what they say about the natural as well as the political environment, this chapter suggests that they depict a specifically postmodern view of nature. This postmodern depiction of the environment is seen as especially interesting for two reasons: for how it addresses the difference in scale between human and environmental time, and for how it questions conventional categorical differentiations between nature and culture, and past and present. Besides Oppermann's study, the analysis builds on Thomas Docherty's postmodern reading of 'The Grauballe Man' in 'Ana-; or Postmodernism, Landscape, Seamus Heaney', extending his analysis of one of the bog poems to carry out a postmodern ecocritical analysis of the bog poems as a group.

Technology and landscape

From abstract postmodern theory Chapter 3 turns to narrative framings of technological progress and its environmental and societal impacts. Focusing on North America in the nineteenth and early twentieth century, David Nye has identified and analysed different categories of narrative responses to technology and environmental change, outlining a simultaneous material and cultural social construction of landscapes, with a 'constant interplay of site and sight'. He points out the crucial role of narratives in this interplay:

> the question of who is making the narrative of a place is just as important as who constructs the physical landscape. Is the Grand Canyon described as a profitless locality that cannot support agriculture, a sublime wonder, or a potential mining site? Is the smoke of an industrial city construed as pollution or a heartening sign of prosperity? Is an exhausted open pit mine filled with water to be understood as a ravaged landscape, an industrial tourist site, or a recreation area? Is the Appalachian Trail a means to revivify the rural economy or a way to escape into a rugged wilderness experience?
>
> (1999: 16, 7–8)

From J.B. Jackson's definition of landscape as 'A composition of man-made or man-modified spaces to serve as infrastructure or background for our collective existence', Nye has also introduced the idea of the 'anti-landscape', a space that no longer sustains any life, human or other (J.B. Jackson, *Discovering the Vernacular Landscape* (1984), quote from Nye, 1999: 3). The relationship between landscapes and anti-landscapes is often complex – the sustenance of one landscape may depend on the creation of an anti-landscape somewhere else, with resources transported to and waste products taken from the privileged area.

Nye identifies three main categories of stories that respond to technological development and landscape: *foundation narratives*, *counter-narratives*, and *recovery narratives*. These different types of narratives frame new technologies and their effects on both human and non-human nature in fundamentally different, sometimes opposing ways. To a certain extent, they reflect ideas and policies of certain times, such as American frontier narratives in the nineteenth century, but they also coexist and co-develop, representing different points of views and frameworks for how to imagine and construct the relationship between social and economic development, human and environmental ethics, the meaning of progress, and so on.

While Nye focuses primarily on European settlers in America in the nineteenth and early twentieth centuries, the poems in Hughes's *Elmet* collections are mostly about the end of that period, and shift our focus from the wilderness frontier in North America to an industrial and post-industrial landscape in Yorkshire in northern England, sometimes portrayed as an anti-landscape. The two *Elmet* collections, from 1979 and 1994 (revised edition), trace the history of the Upper Calder Valley, from the last Ice Age through agricultural cultivation and prosperity during the industrial revolution to post-industrial and post-war decline, with emphasis on industrialisation and thereafter. Drawing on Nye's theoretical framework and narrative categories, this chapter identifies different types of stories and tropes that Hughes's poems evoke and relate to, especially in the form of environmental counter-narratives. Connecting to Nye's 'history of technology'-approach is also an attempt to historicise Hughes's twentieth-century British counter-narratives in relation to nineteenth-century American frontier narratives, placing them in the broader context of the relationship between narratives, technology, and social and natural landscapes.

Postcolonial ecocriticism

As Helen Tiffin and Graham Huggan have noted, concepts such as conquest, indigeneity and native and invasive species, human and other, are central to both postcolonial and environmental studies (Huggan and Tiffin, 2010). Despite this, postcolonial ecocriticism, though suggested early in the history of ecocriticism (see e.g. Buell, 1995: 6), is a relatively recent development in a more substantial sense. Rob Nixon has outlined the surprising, yet persistent division between the two fields:

> These two fields have emerged in recent decades as among the most dynamic areas in literary studies, yet their relationship has been, until very recently, dominated by reciprocal indifferences or mistrust. Unlike many initiatives within literary studies (reader response theory, say, or deconstruction), environmental studies and postcolonial studies have both exhibited an often-activist dimension that connects their priorities to movements for social change. Yet for the most part, a broad silence has characterized environmentalists' stance toward postcolonial literature and theory while postcolonial critics have typically been no less silent on the subject of environmental literature.
>
> (2011: 233)

To a degree, and as Nixon's book indicates, the combination of environmental or ecocritical studies with (post)colonial perspectives is being reframed as studies of environmental justice. However, as Nixon also demonstrates, the colonial framework is still highly relevant and illuminating for studies of environmental relationships between the global north and the global south, highlighting not only uneven effects of environmental change in different parts of the world, but also the extent to which early ecocriticism itself developed as a subfield of American studies, and still favours Anglo-American nature writing traditions (2011: 235). In a response to Nixon, Hedley Twidle notes how, while sharing certain concerns, environmental and postcolonial literary studies have also evolved around opposing ideas and themes: 'If postcolonial theory has concerned itself with ideas of hybridity, cross-cultural mixing, displacement, migrancy and the recovery of silenced histories, then ecocriticism has often been concerned with their opposites: discourses of purity, conservation, regionalism and solitude – moments of timeless communion with nature' (2014: 54).

Linking postcolonial and ecocritical perspectives is a way to consider how both theories and authors considered foundational to traditional ecocriticism may appear from a perspective other than an Anglo-American one. Twidle shows for example how the tropes and rhetoric that make *Silent Spring* so convincing to an American and European audience seem to have a different meaning in a country like South Africa. The silent birds in the title of Carson's book, from a poem by British poet John Keats – 'The sedge is wither'd from the lake / And no birds sing' (from 'La Belle Dame Sans Merci') – have different implications on the campus of University of Cape Town, on the slopes of Table Mountain, from where Twidle is writing, than they do in the English countryside or American suburb:

> birdsong, or its absence, takes on a rather unexpected association in our local habitat. A little further up the slopes of this campus, one finds a statue of the 19th-century mining magnate and empire builder Cecil Rhodes, gazing toward the African hinterland. At one point in his short but energetic life, Rhodes attempted to introduce a variety of songbirds to the slopes of Table Mountain, apparently from a desire to hear the sound of English woodland again before his death. The scheme was unsuccessful – only the chaffinch proved able to 'naturalise' itself – but this was only one element of a much broader programme of green imperialism which makes the Rhodes Estate around us a complex and carefully manufactured cultural landscape. [...] If for Carson birdsong came to symbolise an environment under threat from man's chemical ingenuity and meddling in complex natural systems, then on these slopes it was inflected with the grand designs of settler-colonialism – an index of precisely such an impulse to remake the environment in the image of one's own native land.
>
> (Twidle, 2014: 56)

The presence or absence of birdsong takes on yet another meaning in the (post) colonial context of Heaney's poetry. The fourth chapter of this book examines interrelations between colonial and ecological themes in Heaney's poetics. The setting of these relationships is very different compared to its counterparts in the global south, both ecologically and politically. The geographical proximity between Ireland and the UK means that there are less extreme differences between the natures of the colonisers and the colonised, and this causes a different set of concerns compared to conflicts between geographically distant nations.

While both colonial and ecological themes have been recognised in Heaney's poetics by other critics, political perspectives have taken a clear precedence over environmental readings. Noticing that several poems from Heaney's 1972 collection *Wintering Out* 'register Heaney's awareness of threats posed to his native environments, and in particular the danger of extinction of indigenous species', Greg Garrard has also noted that these environmentally oriented poems 'have failed to draw much attention from critics – not one is discussed in detail in any of the major monographs on Heaney – and have also tended to be excluded from the anthologies' (1999: 193). The lack of interaction between ecocritical and postcolonial analysis is also visible in studies of Heaney – among the many colonial readings of Heaney's poems, so far very few have addressed how many of his poems about language and other colonial themes also engage with ecological concerns, and how social and environmental changes are intertwined with each other.

Focusing on two of Heaney's collections, *Wintering Out* (1972) and *Human Chain* (2010), Chapter 4 demonstrates that ecological themes are not, as some critics have suggested, simply allegorical in Heaney's poems, but a crucial part of his engagement with the relationship between Ireland and the UK. Picking up from Garrard's analysis of 'The Backward Look', which points towards not a complementary ecological interpretation but rather to an 'infolded isomorphism' (1999: 194) between the ecological and colonial themes in the poem, this chapter shows how Heaney intertwines ecological and colonial implications in ways that go far beyond the allegorical.

Ecosemiotics

Initiated by Winfried Nöth and Kalevi Kull in two respective articles in a special issue of *Sign System Studies* in 1998, ecosemiotics forms a bridge between cultural semiotics and biological semiotics, or biosemiotics for short. While cultural semiotics is the study of human sign systems (language, gestures and so on) and biosemiotics studies sign activity in other species (plants or animals or, on a different scale, cells, bacteria or other minute organisms), ecosemiotics takes an interspecies and ecosystemic point of view of different semiospheres and studies, as Nöth suggests in 1998, 'the relationship between the organism and its environment' (1998: 333). Ecosemiotics may study large-scale systems such as forests or oceans, or small-scale systems such as the relationship between different microorganisms.

Kull focuses on ecosemiotics as a means to connect human to non-human sign systems, suggesting that ecosemiotics is perhaps best seen as a subfield to human ecology. He defines ecosemiotics as the study of 'the relationship between culture and external nature':

> *Ecosemiotics* can be defined as the semiotics of relationships between nature and culture. This includes research on the semiotic aspects of the place and role of nature for humans, i.e. what is and what has been the meaning of nature for us, humans, how and in what extent we communicate with nature. Ecosemiotics deals with the semiosis going on between a human and its ecosystem, or a human in one's ecosystem.
>
> (1998: 347)

Birdsong is again an example of how a sign can change meaning not just across different continents, but also across the boundaries of different semiospheres. While it is often a territorial sign in the semiosphere of the bird's own species, humans may interpret the same song as a sign of pastoral tranquillity and peacefulness. By examining how signs travel and are sometimes transformed across species boundaries, ecosemioticians thus study the relationship and communication between different species that share the same ecosystem.

Analysing and building on Kull's and Nöth's definitions of ecosemiotics, Timo Maran, also from the tradition of semiotics at Tartu University, has developed a methodology for ecosemiotic analysis of literary texts that is meant to enable scholars to 'take into account the semioticity of nature itself as well as allow analyzing the depiction of nature in the written texts' (2007: 269). Maran introduces the concept of the 'nature-text' to denote the intertwined meaning-relations between written texts and the natural environment. The 'nature-text' depends both on the imagination and social and cultural context of the writer and on the structures and organisms in the natural world that it describes. Maran uses non-fiction nature writing as the most straightforward example of this unit, but, as the fifth chapter shows, it can also be used to understand ecopoems as a different form of 'nature-text'.

In contrast to most environmentally oriented literary studies, ecosemiotics is an attempt to study representations of and in the natural world from a non-anthropocentric point of view, both in the sense that it highlights sign activity and communication among non-human species, and by trying to understand how environments and environmental change may be experienced by other species in their respective *Umwelten*. The concept of non-anthropocentrism is central in this chapter, where, analysed using a structural ecosemiotic methodology, Hughes's poetry is seen to be anti-anthropocentric in two ways that correspond to the dual anti-anthropocentrism of ecosemiotics as a field; on one hand, the analysis shows how many of Hughes's poems enact what Maran refers to as 'appreciation of an alien semiotic sphere', through poems that focus on non-human sign activity and the semioticity of nature, and on the other hand through poems that attempt to make visible the *Umwelten* of other species (2007: 269). This chapter shows how both these types of poems by Hughes challenge the anthropocentric idea of human exceptionalism.

Globalisation and sense of place

The poem 'The Birthplace' from *Station Island* (1984) asks:

> Everywhere being nowhere,
> who can prove
> one place more than another?

A decade and a half later, Pico Iyer writes in *The Global Soul* that 'Everywhere is so made up of everywhere else – a polycentric anagram – that I hardly notice I'm sitting in a Parisian café just outside Chinatown (in San Francisco), talking to a Mexican-American friend about biculturalism while a Haitian woman stops off to congratulate him on a piece he's just delivered on TV on St. Patrick's Day' (2000: 11). What does this sense of placelessness and geographical interchangeability and interconnectivity mean for the environmental imagination, for a sense of attachment to one's local surroundings and environments?

Ursula Heise examines this question in *Sense of Place and Sense of Planet* (2008). Globalisation, according to Heise, has become the most important organising principle for literary studies, as well as for many other fields:

> Over the last decade and a half, the concept of 'globalization' has emerged as the central term around which theories of current politics, society, and culture in the humanities and social sciences are organized. In literary and cultural studies, it is gradually replacing earlier key concepts in theories of the contemporary such as 'postmodernism' and 'postcolonialism'.
>
> (2008: 4)

In her study, Heise outlines a relationship between developing environmentalism and increasing globalisation, and investigates how these two processes have affected our sense of attachment to different kinds of place, on local and global levels, and how this is reflected in different kinds of aesthetic representations and genres. Heise suggests that, in order to address global environmental concerns, we need new models for how to relate our local surroundings and decisions to their planetary relations and contexts, to help us think locally and globally at the same time.

Heaney's career has unfolded alongside both increasing globalisation and developing environmentalism. This chapter looks for how these different but interrelated processes – of globalisation and deterritorialisation on the one hand and increasing concern for local environments and place attachment on the other – are reflected in his poems. The question of place and place attachment also connects to Heaney as a poet whose work has been translated into multiple languages and includes influences and references from many different cultures, regions and traditions, but at the same time is closely associated with a particular place, and even the actual soil of that place, especially in early poems such as 'Digging'.

The analysis in this chapter focuses on individual poems from different points in Heaney's career, from the 1970s to 2010, tracing a gradual shift from a focus on local place attachment and physically experienced environments to a developing sense of a more abstract idea of a planetary belonging and responsibility. The analysis draws on concepts outlined and developed by Heise, most importantly the ideas of deterritorialisation and (eco)cosmopolitanism, in order to unpack ways in which Heaney's poems suggest how a global and planetary sense of place may be imagined and represented, and how they relate to sometimes conflicting ideas of identity, nationalism and belonging.

By reading Hughes's and Heaney's poems from the variety of thematic and methodological perspectives represented in these chapters this book aims to analyse their work in ways that place their poems in dialogue with disciplines from outside the immediate field of literary and poetry studies, relating to the broader framework of environmental humanities. To allow this kind of conversation and mutual input between poetry analysis and other environmental humanities disciplines, while attention is also paid to the formal qualities of their poems, the main focus in this book is on content, meaning the prevailing themes and ideas, rather than the formal qualities of Heaney's and Hughes's poetics.

Hughes, Heaney and the environment

Both Hughes and Heaney have expressed explicit concerns for the natural environment, in poetry as well as in prose. As several critics have pointed out, Hughes was an early environmental activist and proponent, engaging with environmental issues at least as early as the 1970s. Keith Sagar has noted that in particular after he reviewed conservation biologist Max Nicholson's *The Environmental Revolution* in 1970, Hughes's 'environmental and ecological concerns came to figure more and more centrally both in his poems and in his life, and led to his working for such organisations as the Atlantic Salmon Trust, Farms for City Children and the Sacred Earth Drama Trust (which he founded)' (quote from Gifford, 2009: 142).

In the review of *The Environmental Revolution*, Hughes states that:

> The story of the mind exiled from Nature is the story of Western Man. It is the story of his progressively more desperate search for mechanical and rational and symbolic securities, which will substitute for the spirit-confidence of the Nature he has lost. The basic myth for the ideal Westerner's life is the Quest. The quest for a marriage in the soul or a physical reconquest. The lost life must be captured somehow. It is the story of spiritual romanticism and heroic technological progress. It is a story of decline. When something abandons Nature, or is abandoned by Nature, it has lost touch with its creator, and is called an evolutionary dead-end.
>
> (Hughes, 1994: 129)

This comment outlines and explains the story in *Crow*, for example, published the same year, showing that these were concerns that had occupied Hughes for some time already. It also highlights Hughes's strong belief in an important connection between inner and outer natures; the idea that a 'mechanical' spirit will mistreat nature, while a spiritual attentiveness to the natural world is a crucial part of a harmonious inner life, as suggested in the relationship between technological progress, industrialisation, and human and non-human nature in the *Elmet* poems.

In *The Song of the Earth*, early ecocritic as well as Hughes scholar Jonathan Bate also notes Hughes's concern for the environment, and the significance of his review of Nicholson's book:

> The poetry of Hughes graphically presents a nature that is, in Tennyson's phrase, 'red in tooth and claw'. Yet whilst relishing the violence of nature's own processes, Hughes in his later years became increasingly angry about the violence wrought by man upon nature. He cared deeply about the pollution of the countryside and the decimation of Britain's wildlife population. In 1970, in the first number of a new journal called *Your Environment*, he reviewed *The Environmental Revolution* by Max Nicholson, one of the earliest books to enumerate the full extent of our ecological crisis. Here the future Poet Laureate declared his hand as a card-carrying Green. He wrote of the need to salvage 'all nature from the pressures and oversights of our runaway populations, and from the monstrous anti-Nature that we have created, the now nearly-autonomous Technosphere'.
>
> (2000: 27)

In the same review Hughes also refers to an interconnected universe, where the cells in the body are connected to mice in the field and stars in the sky. Bate points out that Hughes found a sense of both hope and legitimacy in the then 'relatively new science of ecology' (while also noting that Hughes made some spiritual claims regarding these interconnections with which 'A biological scientist would not be entirely happy'), which presented a model of the world where everything is connected to everything else, where the planet is envisioned as a single, fragile unit. As Bate points out, 'For Hughes, this vision binds the ecologist to the poet' (2000: 27, 28).

These environmental concerns are reflected throughout Hughes's poetics, in different shapes and forms. Some poems were composed for a direct, environmentalist purpose. 'The Black Rhino' (Hughes, 2003: 763–7), for example, was written to support a campaign to save the black rhino from extinction in Kenya in the early 1990s:

> This is the Black Rhino, the elastic boulder, coming at a gallop.
> The boulder with a molten core, the animal missile,
> Enlarging towards you. This is him in his fame –

Whose past is Behemoth, sixty million years printing the strata
Whose present is the brain-blink behind a recoiling gunsight
Whose future is a cheap watch shaken in your ear
[…]
For this is the Black Rhino, who vanishes as he approaches
Every second there is less and less of him
By the time he reaches you nothing will remain, maybe, but the horn –
an ornament for a lady's lap

These lines contrast the evolution of the rhino, the slow environmental time of its 'sixty million years' history, with the instantaneity of human time, measured in 'the brain-blink' of a gunshot and the seconds, compared to the sixty million years, it is taking for a few generations to make it extinct. It also contrasts the physical power of the living animal, an 'elastic boulder', with the disposability of the products it is sacrificed to make – 'a cheap watch' or an undefined 'ornament'. Though usually less overt, these contrasts between natural and human time and between animal vivacity and lifeless human objects are a persistent theme in Hughes's poems.

Another Hughesian theme is, as noted, an interconnected universe and humans' recklessness with regard to how they affect it. The poem 'If' (Hughes, 2003: 740) from 1993 exemplifies this theme through the issue of water pollution:

If the sky is infected
The river has to drink it

If earth has a disease which could be fatal
The river has to drink it

If you have infected the sky and the earth
Caught its disease off you – you are the virus
[…]
Already – the drop has returned to the cup
Already you are your ditch, and there you drink.

As Bate notes, 'If' reflects an ecosystemic view of the earth, suggesting that if you pollute or 'infect' one stream or lake, you are to some extent polluting the whole earth. To us in the 2010s this is not a controversial claim, but not so long ago it was. An early proponent of this idea was Rachel Carson, and Hedley Twidle points to a comment in a 1962 review of *Silent Spring* in *Time* magazine that illustrates the newness of this idea at the time; to Carson's statement that you cannot 'add pesticides to water anywhere without threatening the purity of water everywhere', the reviewer replies that 'It takes only a moment of reflection to show that this is nonsense.' As Twidle notes, statements like Carson's now have 'axiomatic status in modern environmentalism' (2014: 52), a status that Hughes also relies on and promotes in 'If'.

The focus on pollution, and primarily the pollution of water, suggests that Hughes was influenced by contemporary concerns that were expressed and debated frequently in the decades leading up to the publication of 'If' in the early 1990s. The poem's description of pollution is especially suggestive of acid rain, describing how 'the sky is infected' by industrial emissions, which rain down on earth and pollute rivers and the drinking water of people, plants and animals. The problem of acid rain was widely recognised among scientists already when Carson was writing in the 1960s, and received significant media attention in the following decades. Hughes refers to it explicitly in 'The Canal's Drowning Black' (Hughes, 2003: 477) from *Remains of Elmet*, published around the same time that the debate about acid rain started to pick up in the late 1970s:[1]

a windowsill
Blackened with acid rain fall-out
From Manchester's rotten lung.

The choice of the term 'fall-out' also alludes to more sinister forms of pollution from nuclear energy, thus pointing not just to the effects of acid rain but to unforeseen consequences of human technologies and industries in a broader sense.

'If' was added to the version of *River* that was included in the 1993 collection *Three Books*, a collection of revised versions of *Remains of Elmet*, *Cave Birds* and *River*. The *River* of this collection includes thirteen additions and nine omissions compared to the original. It also includes a section of notes explaining the background of some of the poems, such as the connection between the health of the river and the fish that live in it, as Hughes expresses his concern for both:

> It is not easy to separate the fascination of rivers from the fascination of fish. Making dams, waterfalls, water-gardens, water-courses, is deeply absorbing play, for most of us, but the results have to be a home for something. When the water is wild, inhabitants are even more important. Streams, rivers, ponds, lakes *without fish* communicate to me one of the ultimate horrors – the poisoning of the wells, death at the source of all that is meant by water. I spent my first eight years beside the West Yorkshire River Calder – in which the only life was a teeming bankside population of brown rats. But the hillside streams and the canal held fish – including, in the canal, big but rare trout. These preoccupied me, as a lifeline might.

The notes go on to record the memory of a lake being poisoned, how suddenly one day Hughes 'saw all the fish in this lake bobbing their mouths at the surface; the beginning of the end':

> That same day I noticed a strange ruddy vein in the ditch water that drained from the farm buildings, two or three hundred yards away. And I registered a

> new smell. I traced the vein to a big stone shed, packed with sodden, dark-stained grass – reeking the new smell. It was the first silage.
>
> (1993: 183–4)

The revisions of *River* for *Three Books* develop the collection's direct engagement with environmental concerns, as the addition of 'If' suggests. Neil Roberts has also noted that these poems are 'more than any of Hughes's earlier poetic work, overtly engaged with ecological crisis', while Terry Gifford describes the *Three Books* version of *River* as 'more radical' than the original (quotes from Gifford, 2009: 57). This development shows an intensifying environmental agenda in Hughes's work, reflecting debates that were picking up in his surroundings.

Not surprisingly against this background, Hughes has been characterised as an 'ecopoet' by several critics. However, most studies of Hughes's ecopoetics so far tend to accept Hughes's own point of view of the relationship between ideas relating to nature, technology and spirituality, which results in analyses that, as Gifford has pointed out, 'combines the ecological with a notion of transcendence' (2009: 141). Gifford mentions Craig Robinson's *The Shepherd of Being* (1989) and Leonard Scigaj's *Ted Hughes* (1991) as examples of this approach. Ann Skea's 'Regeneration in *Remains of Elmet*' (1994) is another example. The discussion of Hughes's poetry in this book is an attempt to analyse not whether or not Hughes can be described as an ecopoet, but rather investigate the particular ideas about nature and environment that are expressed in his poetry, and how these relate to contemporary developments of environmental thought and related interdisciplinary discussions. The analysis thus aims to place Hughes in relation not primarily to a history of nature writing and poetry, but to the emerging field of environmental humanities, taking into account the complexities of environmental change, its implications with regard to relationships between nature and culture, and the broader context of renegotiating ideas about the meaning of the term 'environment'.

In *Stepping Stones*, a collection of interviews with Seamus Heaney by Dennis O'Driscoll, Heaney is asked to what extent his early awareness of environmental issues was influenced by Hughes (and their mutual friend Crooke). Heaney replies:

> Pollution, especially the pollution of rivers, was an obsession with the pair of them, and it was something I myself knew about from childhood. There was always a dread of allowing 'lint water' to get into the Moyola, since it was deadly for the fish – lint water being the water left in a flax dam after the flax had been retted. And I also remembered the sight of the first white froth floating down the Moyola after Nestlé opened their factory at Castledawson. So I was an apt pupil.
>
> (O'Driscoll, 2008: 336)

The change of the river Moyola is reflected in the poem 'Moyulla' (Heaney, 2006: 58–9), from *District and Circle*:

her stones, her purls, her pebbles
slicked and blurred
with algae, as if her name
and addressing water

suffered muddying,
her clear vowels
a great vowel shift,
Moyola to Moyulla.

Elaborating on this poem, Heaney states that 'Moyulla' 'is a praise poem but it's keenly aware of "green" issues', explaining that he 'wanted the darkening of the vowel from "ola" to "ulla" to suggest the darkening of the ecological climate, the pollution of the river over time' (O'Driscoll, 2008: 406). This description of the river's development over time goes even further back, as 'Moyulla' is a rewriting of Heaney's early poem 'A New Song' (Heaney, 1972: 33), from *Wintering Out*. The earlier poem recalls the beauty of the river and its immediate surroundings in its first stanzas:

the river's long swerve,
A kingfisher's blue bolt at dusk
[…]
The Moyola
Pleasuring beneath alder trees

As the poem continues, the recollection of an idyllic past and place is countered by the speaker's memory of a British presence in the area. The vowels that darken in 'Moyulla' to reflect ecological degradation are in this poem juxtaposed with British consonants:

But now our river tongues must rise
From licking native haunts
To flood, with vowelling embrace,
Demesnes staked out in consonants.

Reading 'Moyulla' next to 'A New Song' shows that the ecological meaning of the later poem is also infused with political implications. Both poems provide counter images to the idea of an Irish rural idyll – one through references to colonial oppression and rebellion, the other to environmental degradation. The fourth chapter explores how different these two counter images actually are.

The statement by Heaney in the beginning of this introduction is also taken from *Stepping Stones* (the title of which seems to come from 'A New Song'), and was made in reply to the following question posed by O'Driscoll: 'Do you think poetry can play any practical or meaningful role in changing minds and hearts on environmental issues? In the past you have conceded that no poem is strong enough to stop a tank, so my question is: can a poem stop an SUV?' Heaney replies:

> I think that one answers itself. What has happened, however, is that environmental issues have to a large extent changed the mind of poetry. Again, it's a question of the level of awareness, the horizon of consciousness within which poet and audience can operate. There are those like Gary Snyder and Alice Oswald for whom these matters are an explicit concern, but at this stage nobody can have an uncomplicated Hopkinsian trust in the self-refreshing powers of nature.
>
> (O'Driscoll, 2008: 407)

Greg Garrard has noted that 'It is plain – and has been widely noted (Morrison, Corcoran, Andrews) – that many of Heaney's early poems belong to a counter-pastoral tradition in British poetry that has been most striking in the twentieth century.' Hughes is also part of that tradition; Garrard finds an 'almost overwhelming' influence of Hughes's work on Heaney's earliest collections. Gifford has also discussed this counter-pastoral theme in both Hughes's and Heaney's work in *Green Voices*, from 1995 (second edition 2011), but as Garrard notes, while 'Ecocriticism must concern itself with this counter-pastoral tradition', within it 'only Heaney's least mature work would find a place' (1999: 180, 182). Building on that point, the studies in this book find in fact more differences than similarities between Hughes's and Heaney's ecopoetics. These differences and what they can tell us about the relationship between nature, environment and poetry is discussed in the concluding section of the book.

In an essay titled 'The New You', Anthony Doerr draws attention to our lack of an appropriate language to discuss and conceptualise our relationship with non-human natures and the environments around (and inside) us. He argues that the limited lingual resources offered by concepts such as nature and culture, human and non-human, or inside and outside rely on misguiding assumptions, such as 'that "Me" and "Nature" are discrete entities' for example, and reflect a reality that is much less complex than the actual world (Doerr, 2014). We are nature and vice versa, in a multitude of ways, only one of which is through the millions of microorganisms we carry inside us, one of the examples that Doerr brings up. Inevitably, this lack of a language that enlighteningly and helpfully describes our relationship with other natures makes it difficult to renegotiate that relationship in order to make it more sustainable. The attempts in this book to theorise and discuss Heaney's and Hughes's poetic and environmental imaginations regarding human and non-human natures are meant to frame their poems in ways that enable them to contribute, from different points of view, to this overarching aim of finding a new language for describing and imagining our role and place in relation to the rest of the natural world, and point to how poetry specifically can help us rethink persistent ideas, and start thinking about humans, societies and environments in new ways.

Notes

1 A search for 'acid rain' on Google's 'Ngram Viewer' indicates that the use of the term peaked in 1992, after a dramatic increase that began in the late 1970s (2012-04-25).

References

Bate, J. (2000) *The Song of the Earth*, London: Picador.

Bergthaller, H., Emmet, R., Johns-Putra, A., Kneitz, A., Lidström, S., McCorristine, S., Pérez, I., Phillips, D., Rigby, K., and Robin, L. (2014) 'Mapping Common Ground: Ecocriticism, Environmental History, and the Environmental Humanities', *Environmental Humanities* 5: 261–76.

Buell, L. (1995) *The Environmental Imagination: Thoreau, Nature Writing, and the Formation of American Culture*, Cambridge, MA: Harvard University Press.

Buell, L. (2005) *The Future of Environmental Criticism: Environmental Crisis and Literary Imagination*, Malden, MA: Blackwell Publishing.

Butler, R. (2011) 'Going Green', *Intelligent Life*, Nov./Dec. http://moreintelligentlife.com/content/ideas/robert-butler/going-green [accessed Feb. 2014].

Clark, T. (2011) *The Cambridge Introduction to Literature and the Environment*, New York: Cambridge University Press.

Cole, H. (1997) 'Seamus Heaney, The Art of Poetry No. 75', *The Paris Review*, 144 (Fall). www.theparisreview.org/interviews/1217/the-art-of-poetry-no-75-seamus-heaney [accessed Feb. 2014].

Doerr, A. (2014) 'The New You', *Orion Magazine*, Jan./Feb. www.orionmagazine.org/index.php/articles/article/7971 [accessed Feb. 2014].

Garrard, G. (1999) 'Ecological Literary Criticism: Two Case Studies', doctoral dissertation, University of Liverpool.

Gifford, T. (2009) *Ted Hughes*, London and New York: Routledge.

Heaney, S. (1972) *Wintering Out*, London: Faber & Faber.

Heaney, S. (1984) *Station Island*, London: Faber & Faber.

Heaney, S. (2006) *District and Circle*, London: Faber & Faber.

Heise, U. (2008) *Sense of Place and Sense of Planet: The Environmental Imagination of the Global*, New York: Oxford University Press.

Huggan, G., and Tiffin, H. (2010) *Postcolonial Ecocriticism: Literature, Animals, Environment*, London: Routledge.

Hughes, T. (1993) *Three Books*, London: Faber & Faber.

Hughes, T. (1994) *Winter Pollen: Occasional Prose*, London: Faber & Faber.

Hughes, T. (2003) *Collected Poems*, London: Faber & Faber.

Iyer, P. (2000) *The Global Soul: Jet Lag, Shopping Malls, and the Search for Home*, New York: Vintage Books.

Kull, K. (1998) 'Semiotic Ecology: Different Natures in the Semiosphere', *Sign System Studies*, 26: 344–71.

Maran, T. (2007) 'Towards an Integrated Methodology of Ecosemiotics: The Concept of Nature-Text', *Sign Systems Studies*, 35/1–2: 269–94.

Morton, T. (2007) *Ecology without Nature: Rethinking Environmental Aesthetics*, Cambridge, MA: Harvard University Press.

Nixon, R. (2011) *Slow Violence and the Environmentalism of the Poor*, Cambridge, MA: Harvard University Press.

Nöth, W. (1998) 'Ecosemiotics', *Sign System Studies*, 26: 332–43.

Nye, D.E. (ed.) (1999) *Technologies of Landscape: From Reaping to Recycling*, Amherst, MA: University of Massachusetts Press.

O'Driscoll, D. (2008) *Stepping Stones: Interviews with Seamus Heaney*, London: Faber & Faber.

Oppermann, S. (2006) 'Theorizing Ecocriticism: Toward a Postmodern Ecocritical Practice', *Interdisciplinary Studies in Literature and Environment*, 13/2: 103–28.

Schellnhuber, H.J. (1999) '"Earth System" Analysis and the Second Copernican Revolution', *Nature*, 402 (supp, Dec.): C19–C23.

Steffen, W., Crutzen, P.J., and McNeill, J.R. (2007) 'The Anthropocene: Are Humans Now Overwhelming the Great Forces of Nature?', *Ambio*, 36/8 (Dec.): 614–21.

Taylor, B. (2010) *Dark Green Religion: Nature Spirituality and the Planetary Future*, Berkeley, CA: California University Press.

Twidle, H. (2014) 'Rachel Carson and the Perils of Simplicity: Reading *Silent Spring* from the Global South', *Ariel: A Review of International English Literature*, Special Issue on Postcolonial Ecologies, 44/4: 49–88.

White, L. (1967) 'The Historical Roots of our Ecologic Crisis', *Science*, 155/3767: 1203–7.

1 Ecotrickster

Environment and nature religion in *Crow*[1]

In 1967, three years before *Crow* was published, Lynn White's now classical article 'The Historical Roots of our Ecologic Crisis' appeared in the journal *Science*. White argued that modern technology and science 'are so tinctured with orthodox Christian arrogance toward nature that no solution for our ecological crisis can be expected from them alone'. According to White, Christianity promotes a view of humans as essentially different from the rest of nature, with the right to dominion over the earth and its inhabitants and with the ability, ultimately, to transcend nature. As a result of those views, he suggests, Christianity supports an unsustainable attitude towards the environment. Without addressing this underlying ideology, White concludes that any improvement of the nature–human relationship is difficult, or even impossible: 'Since the roots of our trouble are so largely religious, the remedy must also be essentially religious' (1967: 1207).

White describes Christianity as 'the most anthropocentric religion the world has seen', and argues that the shift from paganism to Christianity signifies 'the greatest psychic revolution in the history of our culture' (1967: 1205). Christian ideology removed spirituality from the natural world and relocated it to an otherworldly realm, a different sphere from our empirical lives. This separation of the empirical from the spiritual realm made it possible to explore and exploit natural resources without taking into account sacred or spiritual entities present in the environment.

In *Dark Green Religion*, Bron Taylor notes that White's article appeared 'at an auspicious cultural moment' of growing receptivity to alternative, non-Western worldviews:

> This period was characterized by growing receptivity to the religious beliefs and practices of indigenous and Asian peoples at the same time that many were rejecting mainstream Western religions. Fused with intensifying environmental alarm, this religion-related ferment provided fertile cultural ground for a robust debate about the relationships between people, religion, and nature.
>
> (2010: 8)

Hughes was aware of this development, and contributed to the debate. In 1980, he described Christianity as 'just another provisional myth':

> There are now quite a few writers about who do not seem to belong spiritually to the Christian civilisation at all. In their world Christianity is just another provisional myth of man's relationship with the creator and the world of spirit. Their world is a continuation or a re-emergence of the pre-Christian world … it is the little pagan religions and cults, the primitive religions from which of course Christianity itself grew.
>
> ('Ted Hughes and Crow: An Interview with Ekbert Faas', in Faas, 1980: 205)

One of these writers was Hughes himself, and this chapter argues that especially his collection *Crow*, published in 1970, is an aesthetic expression of the kind of critique and reconsideration of the relationship between religion and environmental attitudes that was formulated theoretically by White. The story of *Crow* revolves, just like White's argument does, around the relationship between Christianity and science and technology and the implications of that relationship for our view of and relationship with the natural environment. While White formulates a theoretical argument, the *Crow* poems *enact* a similar critique, not outlining but rather staging a different religious mythology.

This chapter reads White's thesis, which sparked lively debate and was challenged by a number of philosophers, historians and theologians (see e.g. Barbour, 1973 and Shaiko, 1987), in tandem with *Crow*, not in order to underline White's argument, but to historicise *Crow* and place it within a contemporary development of a contested relationship between Western culture and environmental degradation.

Crow the trickster

Crow critiques Christianity and other Western worldviews through a trickster narrative. Through deliberate mischief and accidental misunderstandings, the protagonist Crow reveals the separation of the physical world from a spiritual realm as absurd. The second main character in this story, the Christian God, desperately tries to civilise Crow and teach him about the world from a Christian point of view, but is instead repeatedly exposed as illogical, helpless and incapable. By revealing religious-based arrogance towards the natural world and its consequences, the poems counteract the Christian assumption of human superiority and divine protection. Instead, *Crow* suggests an alternative view of humans as ultimately exposed to indifferent and much more powerful forces of nature, promoting an attitude of humility rather than stewardship towards the global ecosystem.

In a foundational study of trickster stories from 1956, Paul Radin defines the trickster narrative as one of the oldest, most widely distributed and least changed of all myths:

> The Trickster myth is found in clearly recognizable form among the simplest aboriginal tribes and among the complex. We encounter it among the ancient Greeks, the Chinese, the Japanese and in the Semitic world. Many of the

> Trickster's traits were perpetuated in the figure of the mediaeval jester, and have survived right up to the present day in the Punch-and-Judy plays and in the clown. Although repeatedly combined with other myths and frequently drastically reorganized and reinterpreted, its basic plot seems always to have succeeded in reasserting itself.
>
> (1956: p. ix)

Radin's characterisation of the American Indian trickster in particular is also an exact description of Crow:

> [he is] at one and the same time creator and destroyer, giver and negator, he who dupes others and who is always duped himself. He wills nothing consciously. At all times he is constrained to behave as he does from impulses over which he has no control. He knows neither good nor evil yet he is responsible for both. He possesses no values, moral or social, is at the mercy of his passions and appetites, yet through his actions all values come into being.
>
> (1956: p. xi)

Several critics have noted the trickster character of Crow. The most thorough analysis of the *Crow* story as a trickster narrative is made by Jarold Ramsey in '*Crow*, or the Trickster Transformed' (1983), where Ramsey recognises Crow as a figure that draws on related characters from traditional folklore from all over the world, including African, Norse, Native American, Greek, Islamic and Hebraic mythologies. Ramsey notes that the story of *Crow* 'follows very faithfully the typical incidents in a cycle of Trickster-narratives', including 'the wild escapades in series, causes leading to improbable effects that snowball in magnitude, maniacal pursuits, villainous transformations, the periodic Bang! that utterly destroys the protagonist, who then appears in the next scene intact, the wholesale inconsistencies between narratives – all this is standard fare in the Trickster story' (1983: 173, 176).

In an ecocritical analysis of the trickster narrative, John Gamber suggests that Native American tricksters 'begin with the understanding that other-than-human elements comprise controlling forces over which they have, and more importantly *should* have, little power'. According to Gamber, the trickster 'not only uses stories to con the people, but is himself a story' – a story that 'operates to liberate' by countering 'multiple levels of confinement, internment, imprisonment, bondage, and limitation', ultimately recreating the world: 'The trickster, though mischievous, is imagined to be innocent; his aim is to recreate the world, to imagine it otherwise' (Gamber, 2009). In *Crow*, the liberating story is threefold: from a Christian worldview, from an exaggerated and unreflexive belief in scientific and technological progress, and from an estranged relationship with the natural world as a spiritual place.

While tricksters are anthropomorphised figures that to some extent are more similar to humans than the animals they appear as, they also draw on the individual mythologies of their specific species. Considering the different genealogies of common trickster animals, especially the fox that plays a central role both in Hughes's collections preceding *Crow* (especially in the poem and background story of 'The Thought-Fox', recounted in Hughes, 1994) and in North American mythology (as a coyote), why did Hughes want the trickster in this collection to be a crow?

In an interdisciplinary study of crows, Boria Sax states that 'No image of an animal is simpler, more iconic, and more unmistakable' than that of the crow. Part of the reason for this, he suggests, is its reputed intelligence:

> Nature writer David Quammen has [suggested] that the natural intelligence of crows is far in excess of what is demanded for survival in their biological niche. The result is that they are continually bored and make up games to amuse themselves.
>
> (2003: 8, 19)

This impression of excess creativity plays a significant role as a prerequisite for much of the mischief that Crow gets up to in the poems. Another possible reason why Hughes chose a crow is their black colour, in the sense that it associates them with dark and mysterious powers, and with blackness as a counter image to the symbol of light in Christian mythology. The colour black plays a central role in Hughes's story. In 'Crow Blacker Than Ever' (Hughes, 2003: 244) Crow flies 'the black flag of himself', while in 'Two Legends' (Hughes, 2003: 217), the very first *Crow* poem, black is the origin of Crow:

> Black was the heart
> Black the liver, black the lungs
> Unable to suck in light
> Black the blood in its loud tunnel
> Black the bowels packed in furnace
> Black too the muscles
> Striving to pull out into the light
> Black the nerves, black the brain
> With its tombed visions
> Black also the soul

The anaphora of the word black in this poem emphasises its symbolic resonances, and the equivocation between that symbolism and an internal darkness creates a sense of a blackness that stretches from the relatively tangible reality of Crow's insides to a subjective or fantasy realm suggested by the reference to a 'black rainbow'.

Another of Crow's typical trickster characteristics is his appetite: Sax notes that especially in its crow or raven form, the Native American trickster is 'notorious for his voracious appetite'. The indiscriminate appetite of the trickster is illustrated by this Hopi story, summarised by Sax:

> a crow once invited his friend the hawk to dinner. Though the fastidious raptor would eat only freshly killed meat, the crow served him a greasy bullsnake that had already begun to decay. The hawk politely pretended to eat and even complimented the crow on his culinary art, while secretly plotting revenge. Soon afterwards, the hawk invited the crow to dinner, and he served a putrid dish concocted from the skin and entrails of rabbits. Instead of turning away in disgust, the crow avidly devoured the meal, leaving the hawk more infuriated than ever.
>
> (2003: 92, 98–9)

This story is so reminiscent of Hughes's Crow that it could be one of the poems in his story. Many of the *Crow* poems similarly revolve around Crow's sometimes just instinctive, sometimes compulsive appetite.

Crow also exhibits the typical ability of tricksters to survive the most impossible situations, sometimes through luck, sometimes through metamorphosis, and sometimes even through reincarnation, in ways that bring to mind the plasmatic nature of animated characters (see e.g. Heise, 2014). Like other raven tricksters, Crow is 'a quintessential survivor' (Sax, 2003: 96), and as spelled out in 'Examination at the Womb-door' (Hughes, 2003: 218–19), he is in fact 'stronger than death':

> Who is stronger than hope? *Death*
> Who is stronger than the will? *Death*
> Stronger than love? *Death*
> Stronger than life? *Death*
>
> But who is stronger than death?
> *Me, evidently.*

Sax quotes these lines in his book about crows, and suggests that for Hughes the cleverness of crows is a sign of 'infernal magic', describing Crow as a protagonist that is 'voracious and pitiless yet indestructible', who, though sometimes defeated, somehow 'always manages to survive' (2003: 153, 154).

Hughes himself explains the choice of Crow as his lead character by referring to folktales wherein

> the prince going on the adventure comes to the stable full of beautiful horses and he needs a horse for the next stage and the King's daughter advises him to take none of the beautiful horses that he'll be offered but to choose the dirty, scabby little foal. You see, I throw out the eagles and choose Crow.
>
> (Quoted from Faas, 1980: 208)

With its origin in folklore and indigenous cultures, the trickster narrative counteracts Christian narrative and logic, and represents a comic rather than a tragic worldview. Compared to Christian images of divine love and selfless acts of sacrifice and forgiveness, the trickster story is recognised, as Radin points out, by

'Laughter, humour and irony' (1956: p. x). *Crow* thus not only offers a critique of Christianity in the specific episodes described in the poems, but likewise through the narrative structures and devices underlying the *Crow* collection as a whole.

Crow and the environment

The underlying assumption of *Crow* is that humanity is pushing the earth towards some kind of collapse. In this sense it is an apocalyptic narrative, as seen for instance in the poem 'A Disaster' (Hughes, 2003: 226–7). In this poem, a 'word' comes and threatens the existence of both people and planet. The 'word' symbolises the human ability to think abstractly, making possible first and foremost the development of language, but also the connected development of science, religions and other ideas. The poem describes how this 'word' turns on its creator like a kind of Frankenstein's monster, causing damage both to the natural world and the built environment:

> There came news of a word.
> Crow saw it killing men. He ate well.
> He saw it bulldozing
> Whole cities to rubble. Again he ate well.
> He saw its excreta poisoning seas.
> He became watchful.
> He saw its breath burning whole lands
> To dusty char.
> He flew clear and peered.

At the end of the poem, the earth survives but people do not. Crow looks on while the word is

> drinking out all the people
> till there were none left,
> all digested inside the word.
>
> Ravenous, the word tried its great lips
> on the earth's bulge, like a giant lamprey –
> there it started to suck.
>
> But its effort weakened.
> It could digest nothing but people.

And so, 'Its era was over'. The poem develops the story of the fall of man, as it describes how knowledge, materialised in the form of language and represented by the word, gives rise to different ideas and developments that eventually turn on the people that have created it, and destroys them. The word can be read as representing particular ideas or dogmas, that run their course and are then replaced by others, but it can also be read as representing human language in a more encompassing manner, in which case the era that is over at the end of the poem is the era of humans.

In 'Revenge Fable' (Hughes, 2003: 244–5), further on in the collection, both planet and people are destroyed. In this poem, 'mother' refers to mother nature and the 'person' represents people. The story tells how, in the process of trying to fight themselves free of the natural world, people lose sight of their dependence on that same earth:

> There was a person
> Could not get rid of his mother
> As if he were her topmost twig.
> So he pounded and hacked at her
> With numbers and equations and laws
> Which he invented and called truth.
> [...]
> With all her babes in her arms, in ghostly weepings,
> She died.
>
> His head fell off like a leaf.

Different from 'A Disaster', in which perhaps the era of humanity ends but the earth survives, in this poem both people and planet perish.

Hughes's portrayal of destruction in these poems is partly influenced by William Blake, a recurring influence in Hughes's work. In *Ted Hughes and Nature*, Sagar points out that Hughes shares with Blake a view of the condition of Western man as recognised by a singularity of vision. According to Sagar, Blake saw four symptoms of this singular vision: 'the deification of reason and the five senses (Locke), mechanistic science (Newton), the increasingly repressive Puritanism of the churches, and the first mills of the Industrial Revolution'. These four reasons are recalled in the *Elmet* poems (discussed in Chapter 3), as is their opposite, Blake's fourfold vision pinnacled by what Sagar describes as 'a vision of the holiness of everything that lives' (2009: 3, 52).

Both 'A Disaster' and 'Revenge Fable' suggest that humanity's increasing separation from the natural world will lead to its demise. Similar images of disaster inform one of the most common tropes in environmentally oriented texts, namely the apocalypse. Greg Garrard notes the long history of this trope:

> it seems likely that the distinctive construction of apocalyptic narratives that inflects much environmentalism today began around 1200 BCE, in the thought of the Iranian prophet Zoroaster, or Zarathustra. Notions of the world's gradual decline were widespread in ancient civilisations, but Zoroaster bequeathed to Jewish, Christian and later secular models of history a sense of urgency about the demise of the world.
>
> (2012: 93)

Further elaborating on this trope, Garrard refers to Stephen O'Leary's characterisation of apocalyptic stories as either comic or tragic. O'Leary states that, while tragedy is primarily informed by the notion of guilt, the mechanism of comedy 'is recognition rather than victimage, and its plot moves not toward sacrifice but to the exposure of fallibility'. Garrard adds that, while tragic time is 'predetermined and epochal, always careering towards some final, catastrophic conclusion, comic time is open-ended and episodic'. Within the comic frame, 'Human agency is real but flawed', with individual characters who 'are typically morally conflicted and ambiguous' (Garrard, 2012: 95).

This also describes the *Crow* story, in which the individual poems form different episodes, humans are continually mistaken about themselves as well as about the world, and Crow, though often without a moral framework, in some poems can be seen struggling with ethical considerations of his own behaviour (in 'Crow Tyrannosaurus', for example). *Crow* is thus a comic-apocalyptic narrative (in addition to its comic nature as a trickster story). While some of the individual poems, like 'A Disaster' and 'Revenge Fable', can be described as tragic and 'careering towards some final, catastrophic conclusion', these poems are not representative of the collection as a whole, as those individual poems are contradicted by succeeding poems in which both the world and Crow appear restored.

While *Crow* adopts parts of the apocalyptic trope in its narrative structure, it also critiques that same trope based on the fact that images of apocalypse are to a significant degree informed by a Christian tradition, as Garrard has explained:

> Orthodox, Roman Catholic and, for the most part, Protestant Christianity has promoted comic apocalypticism. The imperatives of scriptural authority, history and popular enthusiasm have rendered the trope indispensable, but a tragic frame tends to produce either schisms or perpetual charismatic revolution, and seems unsustainable in the long term. The implications for attitudes to the natural world, moreover, seem worse in the tragic mode. We may recall Lynn White Jr's argument that Christianity is a dangerously anthropocentric religion, and perhaps his parenthetical comment that only Zoroastrianism might be comparable to it. White draws attention to the dualistic conception of humanity and nature that the two religions share, but in addition they are both apocalyptic, which may be the key to the question of Judaeo-Christianity's contribution to environmental problems.
>
> (2012: 96)

By adopting elements of apocalypse, the *Crow* story is not always in strict opposition to the Christian narrative, but sometimes appears intertwined with it. This suggests that *Crow*, while critical of Christianity, does not free itself entirely from elements of Christian images. Ramsey notes this tendency when he argues that *Crow* cannot really be described as representing 'a "post-Christian view"', since even though

> the central impulse of the *Crow* poems is certainly an ingenious and unrelenting subversion of the Christian mythos, so as to reveal how it has got nearly everything wrong about Man's origins [...] the violence of the subversion, the sense of overkill, in fact, indicates that for the author Christianity is still much more immediate and formidable that [sic] 'just another provisional myth'. Better to say that in these poems, Hughes tries to fight himself free of a still-prevailing Christian-humanistic frame of reference that in its omissions and distortions of human facts makes our inherently bad lot a good deal worse.
> (1983: 173)

Hughes's use of images of apocalypse supports Ramsey's suggestion that elements of a Christian myth still sometimes prevail in *Crow*.

In his article, White estimates that the view of the natural environment as spiritual has been marginalised roughly since around the time of the Middle Ages. Since then humans have gradually gained knowledge about and control over the natural world, and acquired sophisticated tools with which to modify and manipulate it. In 'Crow and Mama' (Hughes, 2003: 219–20), this development is described as humanity's imagined liberation from the earth, where Crow represents humanity while the earth is his 'mother', again using the idea of 'mother earth':

> When Crow cried his mother's ear
> Scorched to a stump.
>
> When he laughed she wept
> Blood her breasts her palms her brow all wept blood.
>
> He tried a step, then a step, and again a step –
> Everyone scarred her face for ever

Crow moves further and further away from his mother, or at least so he thinks. With an eye to the space race culminating in the first moon landing, taking place just one year before the publication of this collection, Crow even gets into a rocket in order to get away from his mother/the earth, even though his 'trajectory / Drilled clean through her heart'. In the rocket he finally feels 'cosy', due partly at least to the fact that he does 'not see much': he can see only parts of the earth through the small portholes in the rocket, a reference perhaps to a limited or specialised view of the world represented by science and technology. To his shock, however, when the rocket eventually crashes on the moon, Crow finds himself crawling 'out / Under his mother's buttocks'. The poem points to the fact that no matter how removed humanity might think it is from the natural world, it is in fact not detached from it at all. It also draws a parallel to how one of the most significant outcomes of the expedition to the moon is the images of the earth as seen from outer space it enabled, especially the 'blue marble' photograph, which quickly became a symbol for numerous environmentalist movements (the significance of this image is discussed in relation to Heaney's poems in Chapter 6 of this book).

Christianity

Like White, *Crow* seems to suggest that Christianity has contributed more than any other cultural or ideological framework to the estrangement of humanity from the natural world. The view of man as the final and finest of God's creations, superior to all other species and with the right to dominion over the earth, results in man's view of himself as separated from rather than part of the natural world. The Christian creation story, which provides the basis for this view of man, is altered, but not beyond recognition, in the second poem in *Crow*, titled 'Lineage' (Hughes, 2003: 218). This poem draws from the genealogies in Genesis as well as from John 1: 1 to set the stage for the ensuing story about the adventures of Crow. The biblical statement that 'In the beginning was the word' is replaced by 'In the beginning was scream'. The scream then begets a long list of things, including 'blood', 'fear', 'bone' and 'granite' before it begets Adam and then Mary, and, tellingly, only then does it beget God, who then begets 'nothing'. Then there is a long break of nothing and never, before Crow appears:

> Who begat Adam
> Who begat Mary
> Who begat God
> Who begat Nothing
> Who begat Never
> Never Never Never
>
> Who begat Crow

This account of creation replaces civilised Christian concepts with their primal equivalents, most clearly replacing 'word' with 'scream'. It also reverses the biblical order of events so that the appearance of God comes *after* that of humans, thus suggesting that God appears as a product of human imaginations.

As Ramsey notes, 'Lineage' is a 'mockery of Biblical genealogies' (1983: 178). Another enactment of this theme occurs in 'A Horrible Religious Error' (Hughes, 2003: 231), which rewrites the story of the fall from Eden. When the serpent emerges, 'earth-bowel brown / From the hatched atom', God, Adam and Eve are paralysed by fear, their knees trembling, and they whisper to the snake: 'Your will is our peace'. Crow, on the other hand:

> only peered.
> Then took a step or two forward,
> Grabbed this creature by the slackskin nape,
>
> Beat the hell out of it, and ate it.

The fact that the serpent emerges from a 'hatched atom' rather than a hatched egg updates the biblical story of the tree of knowledge with references to contemporary science; the hatched atom seems to be a reference to nuclear fission and the risks of both radioactive fall-out and nuclear war, two concerns on political agendas in the decade leading up to the publication of *Crow*.

Further pondering the relationship between God and man, in 'Crow's Theology' (Hughes, 2003: 227) Crow first thinks that he owes his existence to God, realising that 'God loved him – / Otherwise he would have dropped dead'. It's a pleasant and reassuring thought: 'Crow reclined, marvelling, on his heart-beat'. However, as Crow continues this train of thought, increasingly difficult questions appear:

But what
Loved the stones and spoke stone?
They seemed to exist too. –

These questions become more serious, as Crow begins to consider death in different forms, his own as well as other crows':

And what spoke that strange silence
After his clamour of caws faded?

And what loved the shot pellets
That dribbled from those strung-up mummifying crows?
What spoke the silence of the lead?

His conclusion is that there are 'two Gods':

One of them much bigger than the other
Loving his enemies
And having all the weapons.

The 'much bigger' god in this poem is the natural world, which is both indifferent and more powerful than the Christian God. The distinction between these two entities suggests the Gnostic notion that the Creator is a demiurge of quite limited power, distinct from the true deity that exists completely aloof from, and contemptuous of, the human world. Unlike the disabled but caring Christian God, this biocentric God is mighty (it has 'all the weapons') and threatening ('loving his enemies'). It does not care for anyone or anything in particular, and it makes no distinction between humans, crows, stones, lead or silence. It does not care or take notice about individual lives, but only cares about the principles of the creation, the ongoing cycle of life and death.

Throughout the *Crow* collection, several important Christian concepts are similarly questioned and ridiculed. In 'Crow's First Lesson' (Hughes, 2003: 211), God tries to teach Crow to say the word 'love'. Crow, of course, fails dramatically and excessively: instead of pronouncing the word 'love', he retches, gags and produces first a shark and then, in quick succession, 'a bluefly, a tsetse, a mosquito', a group of disease transmitting insects, followed by 'Man's bodiless prodigious head', and finally 'woman's vulva'. The poem is a catastrophic failure on God's part; instead of educating Crow he is reduced to tears and cursing as he tries to separate the various body parts that have begun fighting in front of him. Crow, unreformed, flies 'guiltily off'. The poem is among other things a critique of theodicy, the attempt to account for God's justice and the problem of natural evil, as the Christian concept of 'love' is replaced by a number of counter images of violence, disease and fighting.

A similar failure occurs in 'Crow Communes' (Hughes, 2003: 224). In this poem, Crow sits on a mountain that is also God's shoulder. God, meanwhile, lies 'agape, a great carcase', paralysed by Crow who has taken the form of a literal 'chip on his shoulder'. The word 'agape' suggests an image of God as gaping, stupefied. The origin of the word agape, however, is 'agapan', meaning love or affection, especially in the sense of a selfless and self-sacrificing Christian love. It is also connected to early Christian love feasts, where meals were eaten in the name of Christ. 'Agape', against this background, evokes God's love of humanity, as well as a more general Christian love. By pairing 'agape' with 'carcase', and by previous adjectives referring to God as 'exhausted' and 'snoring', the poem associates this meaning of 'agape' with an image of God as debilitated and selfless to the point of complete powerlessness and incapacitation.

Not surprisingly, the religious concept of agape is entirely lost on Crow, as is that of communion. Playing the part of 'hierophant', meaning priest or interpreter, Crow interprets communion and the idea of the Christian feast by tearing off an actual piece of God's shoulder. He eats it and confirms that 'it's true, he suddenly felt much stronger'. The misunderstanding, the turning of an abstract concept into physical and empirical reality, is similar to the events in 'Crow's First Lesson'.

In 'Crow Tyrannosaurus', Crow reflects on the idea of Christian morality, and is overcome by what he suddenly perceives to be the horrors of the food chain:

> It was a cortege
> Of mourning and lament
> Crow could hear and he looked around fearfully.
>
> The swift's body fled past
> Pulsating
> With insects
> And their anguish, all it had eaten.

Crow wonders if he should try to change his ways:

> 'Alas
> Alas ought I
> To stop eating
> And try to become the light?'

But of course he is unable to. His evolutionary predisposition outweighs his moral doubts and he is trapped by his instincts. The struggle between his instinctual and moral selves becomes another factor that further drives the evolution of his being, as Crow becomes all crows:

> But his eye saw a grub. And his head, trapsprung, stabbed.
> [...]
>
> Weeping he walked and stabbed
>
> Thus came the eye's
>   roundness
>   the ear's
>   deafness.

The contrast between Crow trying 'to become the light' and the evolution of deafness so as not to hear the weeping of his prey shows a Christian worldview in conflict with a Darwinian natural order, reflecting, as Ramsey notes, man's 'predicament as conscious beast, human animal' (1983: 180). The conflict stays with Crow for several poems, as he struggles to find a way to reconcile it.

Pastoral nature

In relation to its critique of Christian ideology, *Crow* also reproves a Romantic, pastoral view of nature. The pastoral tradition, perhaps especially in poetry, typically portrays a natural world that is idyllic, pristine and peaceful, as opposed to violent, predatory and unpredictable. In *Crow*, this pastoral view of nature is both deceptive and perilous, as imagined in the poem 'Glimpse' (Hughes, 2003: 256). In this poem, Crow, suddenly possessed by the spirit of Romanticism, starts to serenade the trees, with disastrous results:

> 'O leaves,' Crow sang, trembling, 'O leaves –'
>
> The touch of a leaf's edge at his throat
> Guillotined further comment.

Thus, Crow is given a lesson about the dangers of disregarding or romanticising unpredictable and potentially deadly forces of nature.

Even so, Crow fails to learn his lesson, and 'Glimpse' concludes with Crow continuing 'to stare at the leaves / Through the god's head instantly substituted'. The substitution signifies a change from one set of false beliefs to another, rather than from falsehood to truth. The poem thus critiques not only a pastoral view of nature, but the tendency to relate to nature through any sort of falsifying frame, be it pastoral, Christian or other. The glimpse in the poem's title is perhaps a reference to the instant between two such frames.

As a trickster figure, Crow is in himself an anti-pastoral manifestation. The poem 'Crow and the Birds' (Hughes, 2003: 210), about Crow's indiscriminate appetite, tells its story partly through the change in language that occurs between the beginning and the end of the poem, suggesting a move from pastoral to anti-pastoral.

When the eagle soared clear through a dawn distilling of emerald
When the curlew trawled in seadusk through a chime of wineglasses
When the swallow swooped through a woman's song in a cavern
And the swift flicked through the breath of a violet
[...]
Crow spraddled head-down in the beach-garbage, guzzling a dropped ice-cream.

The change from pastoral-sounding 'emerald', 'dawn', 'seadusk' and 'breath of a violet' in the first stanza to 'spraddled head-down' and 'guzzling' in the last line emphasises Crow's anti-pastoral character. The poem also provides Crow with a different sense of historical setting compared to the other birds; while they seem to exist in an almost timeless, perhaps medieval world, Crow appears in close proximity to the present, a setting that also contributes to his anti-pastoral, non-Romantic character.

Science and technology

The most important target for critique in *Crow* after (and in relation to) Christianity is science and technology. Where Christianity allows people to hide behind the illusion of a benevolent God, science, according to *Crow*, allows people to hide behind numbers. The scientific or technological determinism criticised in *Crow* makes people slaves under their own inventions, most directly addressed in 'Crow's Account of the Battle' (Hughes, 2003: 222–3). This poem describes a war, perhaps the Second World War, where soldiers kill each other with increasing efficiency:

The cartridges were banging off, as planned.
The fingers were keeping things going
According to excitement and orders.
The unhurt eyes were full of deadliness.
The bullets pursued their courses
Through clods of stone, earth and skin,
Through intestines, pocket-books, brains, hair, teeth
According to Universal laws.
And mouths cried 'Mamma'
From sudden traps of calculus,
Theorems wrenched men in two,
Shock-severed eyes watched blood

Squandering from a drain pipe
Into the blanks between the stars.

These lines suggest acts of violence that are carried out not according to human decisions, but according to 'Universal laws' of maths and physics. Decisions are made elsewhere, detached from the immediate context where the violence takes place. The science and technology that make these acts possible are the same as that of 'the hatched atom' from which the serpent emerges in 'A Horrible Religious Error'. In 'Crow's Account of the Battle', the combination of science and technology with religious dogma amplifies devastation:

Reality was giving its lesson,
Its mishmash of scripture and physics,
With here, brains in hands, for example,
And there, legs in a treetop.

In the end there is 'no escape except into death'. Once 'the explosives ran out' and 'the smoke cleared', though at first 'everybody wept, / or sat, too exhausted to weep', soon it became clear that 'This had happened too often before / And was going to happen too often in the future'. The story will recur because people and their memories fade too easily:

Bones were too like lath and twigs
Blood was too like water
Cries were too like silence
The most terrible grimaces too like footprints in mud

Rand Brandes suggests that 'while there are many mini-crises throughout [*Crow*], the main ones are encapsulated in "Crow's Account of the Battle"':

> Theorems, scripture and physics lead the self away from the instincts that make us healthy and whole. They separate us from divine creation and our natural spiritual needs. Over time and in isolation they produce a desensitized and fragmented self and society capable of unimaginable atrocities.
>
> (2011: 73)

White argues in his article that the combination of science, technology and religion has led to increasing disregard for the natural world. In 'Crow's Account of the Battle', Hughes widens this disregard to include not just the natural environment, but human nature as well, suggesting that increasing reliance on science and technology as framed by Christian ideology leads to a separation of the human spirit from the natural world, a separation that in turn makes humans uncompassionate and, as Brandes suggests, desensitised to the effects of their technologies.

The combination of science and technology in *Crow* does not only harm humans, it also devastates the environment, as seen in poems like 'Crow and Mama' and 'A Disaster'. White states that contemporary responses to environmental problems 'seem too partial, palliative, negative: ban the bomb, tear down the billboards, give the Hindus contraceptives and tell them to eat their sacred cows'. Though White is not able to provide a better solution, he suggests that we begin 'by looking, in some historical depth, at the presuppositions that underlie modern technology and science'. He identifies the introduction of a new plough in northern Europe in the seventh century as a pivotal moment in this history. While the new plough was much more efficient than its predecessor, it required more oxen than one household normally owned. In order to accommodate cooperation between different families, fields were redistributed. According to White this meant that

> distribution of land was based no longer on the needs of a family but, rather, on the capacity of a power machine to till the earth. Man's relation to the soil was profoundly changed. Formerly man had been part of nature; now he was the exploiter of nature. Nowhere else in the world did farmers develop any analogous agricultural implement. Is it coincidence that modern technology, with its ruthlessness toward nature, has so largely been produced by descendants of these peasants of Northern Europe?
>
> (1967: 1204)

White's analysis places the crucial moment when the relationship between humans and the natural world started to change much further back in history than most other analysts in this field; contemporary climate scientists often suggest that this change began with the industrial revolution (e.g. Crutzen and Stoermer 2000), while Rachel Carson for example suggested that significant changes took place as late as the mid-twentieth century (see Carson, 1962).

According to White, the ruthlessness towards nature that he sees as initiated in the seventh century was legitimised by a simultaneous transition from paganism to Christianity. According to pagan animism, nature was imbued with and protected by different spirits – before you could cut down a tree or mine a mountain you had to placate the spirit of that particular tree or mountain. By replacing this animistic view of nature with a belief-system that removed spirits from the natural world to an imagined heaven, Christianity, according to White, 'made it possible to exploit nature in a mood of indifference' (1967: 1205).

White's argument sparked instant debate among Judaeo-Christian theologians in the late 1960s and early 1970s. Already in 1973 an article was published that provided an overview of the articles that had by then appeared in direct response to White (Rogers, 1973). Most responses took one of three forms: urging a different translation of the word 'dominion' in Genesis; identifying notions relevant to earth care in the Bible beyond Genesis; or emphasising the adaptability of Christian theology in response to new conditions regardless of its influence in the past. The field of eco-theology has since continued to develop, including for example studies such as Rosemary Radford Ruether's *Gaia and God: An Ecofeminist Theology of Earth Healing* (1992), an ambitious study that brings a feminist perspective to the analysis of the relationship between Western culture, Christianity, ecology and environmentalism, and Lionel Kochan's *Beyond the Graven Image: A Jewish View* (1997), which examines the relationship between Judaism, materiality and animism. The extent to which studies in this field oppose or agree with White varies, but they have in common the view that religious and spiritual beliefs profoundly influence our relationship with the natural environment, and therefore need to be addressed as part of the development of a more sustainable relationship between people and planet. Bron Taylor's study *Dark Green Religion*, discussed in the last section of this chapter, is another contribution to this field.

In *Crow*, the marginalisation and pushing aside of a primitive, uncivilised nature spirit by scientific and technological 'progress' is depicted in 'Crow's Undersong' (Hughes, 2003: 237), which tells the story of how nature ('she') attempts but fails to enter a Western, civilised world. Unable to handle modern tools, 'she cannot come all the way':

> She comes singing she cannot manage an instrument
> She comes too cold afraid of clothes
> [...]
> She comes dumb she cannot manage words

In spite of these inabilities, nature is 'amorous' and brings hope, as without 'hope she would not have come'. Without her, there would be 'no crying' and 'no city', as without her there would be nothing at all. Terry Gifford notes that 'Crow's Undersong' signifies 'Crow's attempt to envision the goddess of nature', that the poem 'is a celebration of all that remains of a raw force that is now "under" the trappings of civilisation and conscious, rational life' (2009: 43).

An alternative view of the nature–human relationship

Towards the end of *Crow*, some poems, and especially the very last poem, 'Littleblood', attempt to relocate and restore the kind of animistic nature spirit described both in White's article and in 'Crow's Undersong' as having been pushed aside. These poems are preceded by several poems in which Crow tries but fails to find alternatives to a Christian framework for relating to the natural environment. In 'Crow and the Sea' (Hughes, 2003: 252), for example, Crow fails to understand or relate to the sea in any meaningful way, as the sea is just too vast for him to grasp:

> He tried ignoring the sea
> But it was bigger than death, just as it was bigger than life.
>
> He tried talking to the sea
> But his brain shuttered and his eyes winced from it as from open flame.

[...]
He tried just being in the same world as the sea
But his lungs were not deep enough

And his cheery blood banged off it
Like a water-drop off a hot stove.

Finally he gives up, turns his back and marches 'away from the sea / As a crucified man cannot move'. What Crow does not realise when he turns and walks away from the sea is that he is inevitably walking towards a different shore. The image illustrates how the sea, representing the natural, non-human world, is larger than Crow can apprehend. From this insight come feelings of insignificance, of his lungs being 'not deep enough'.

If 'Crow and the Sea' illustrates the earth's incomprehensible vastness, 'Crow's Nerve Fails' (Hughes, 2003: 232) suggests that it is nevertheless finite and vulnerable. Crow, stricken by guilt as he looks back over his own history, realises that 'His prison is the earth', and that his own prosperity has been at the expense of others. As he reflects on everything he has eaten, he is, as in 'Crow Tyrannosaurus', horrified as he 'Finds his every feather the fossil of a murder'. He wonders:

Who murdered all these?
These living dead, that root in his nerves and his blood
Till he is visibly black?

These lines show an emerging sense of morality in Crow's usually instinct-driven brain. While he notes the inevitability of the rules of his being, asking: 'How can he fly from his feathers?', at the same time he cannot escape a sense of guilt, again struggling with the idea of ethical responsibility and compassion in combination with a Darwinian worldview:

He cannot be forgiven.
His prison is the earth. Clothed in his conviction,
Trying to remember his crimes

Heavily he flies.

While 'Crow's Nerve Fails' is a development of 'Crow Tyrannosaurus', 'Crow Frowns' (Hughes, 2003: 234) represents the next step in Crow's thinking about the relationship between a moral and natural order. In 'Crow Frowns', Crow's adherence to the inevitable rules of his being is a source of freedom rather than guilt. Here, instead of guilt, Crow's 'eating is the wind', suggesting that he feels like a part of nature; his actions no more immoral (or moral) than those of the elements. The poem, and the negotiation that includes the two previous poems as well, concludes with a sense of wonder rather than horror at the process of evolution:

> We are here, we are here.
> He is the long waiting for something
> To use him for some everything
> Having so carefully made him
> Of nothing.

These lines are a sharp contrast to the weeping and stabbing in 'Crow Tyrannosaurus', and show Crow moving towards an acceptance of himself as part of the natural world, without feelings of guilt for how his behaviour has evolved.

In 'Crow Goes Hunting' (Hughes, 2003: 236), the word from 'A Disaster' returns, this time in the form of a group of words, a hunting party ('a lovely pack'). The words are sent out by Crow to catch a fleeing hare, this poem's representative of nature. As it turns out, no matter how cunning the words and Crow are, the hare is able to parry each of their attacks with superior defensive tricks. The hare sets off the story as, with the pack of words on its tail, it

> converted itself to a concrete bunker.
> The words circled protesting, resounding.
>
> Crow turned the words into bombs – they blasted the bunker.
> The bits of bunker flew up – a flock of starlings.
>
> Crow turned the words into shotguns, they shot down the starlings.
> The falling starlings turned to a cloudburst.
>
> Crow turned the words into a reservoir, collecting the water.
> The water turned into an earthquake, swallowing the reservoir.

Turning back to its original form, the hare 'leaped for the hill':

> Crow gazed after the bounding hare
> Speechless with admiration.

Crow enters the hunt full of confidence in his 'well-trained' words with 'strong teeth', dismissing his opponent with 'what is a hare?' – as it turns out, the hare is capable of outmanoeuvring each of Crow's and his words' tricks; no human invention thought of by Crow can outmatch the forms of nature accessible to the hare – it has at its disposal all natural elements and phenomena, including rain, earthquakes and other animals. The poem recalls the story of Proteus, a sea god who can tell the future but who changes shape in order to avoid being captured and forced to do so – his name reappears in the adjective 'protean', meaning versatile or adaptable. The story is told by Ovid in 'The Changes of Proteus' in *Metamorphoses*, a narrative of nature and animals that, just like *Crow*, formulates an alternative creation story to Genesis.

Before the hare leaps for the hills he eats Crow's words. 'Crow Goes Hunting' thus reverses the outcome in 'A Disaster' where the word attempts (but fails) to swallow the earth. The last line of 'Crow Goes Hunting' in which Crow is 'Speechless with admiration' evokes a sense of wonder similar to that in 'Crow Frowns'.

'Crow and the Sea', 'Crow's Nerve Fails', 'Crow Frowns' and 'Crow Goes Hunting' show Crow trying to understand the natural world without reference to a Christian, pastoral or technological deterministic framework. Instead these poems try to imagine alternative, mystical or magical views of the world, with origins other than Western. Rand Brandes observes that the trickster narrative is only one of several mythical frames that can be traced in *Crow*:

> Hughes knew from the very beginning that the vast array of myths, symbols and magical arts that he absorbed from anthropological sources from around the world would define his work and defy his audience. He enhanced these anthropological sources over time by adding a full range of materials from the occult sciences and esoteric philosophies he found in poets like W. B. Yeats and Robert Graves in addition to the depth psychology of Carl Jung and Mircea Eliade's work on comparative religion. The typically ancient and often non-Western sources that these writers mined shaped what he wrote, how he wrote and, most importantly, why he wrote. Hughes, as mythic poet, wrote to liberate and heal – the soul, the body, the mind, the community and the world.
>
> (2011: 68)

The mythical framework that Hughes creates and employs can be characterised by the term 'dark green religion', defined by Bron Taylor in a study with the same title. Taylor describes dark green religion as a 'religion that considers nature to be sacred, imbued with intrinsic value, and worthy of reverent care' (2010: p. ix). While Hughes's version of the inherent worth of nature is held in tension with a Darwinian worldview, the latter is in the end included rather than seen as opposed to a sacred natural world. Taylor also defines dark green religion as, without formal texts or institutions, 'reinforced and spread through artistic forms that often resemble, and are sometimes explicitly designed, as religious rituals'. These rituals, moreover, seek 'to destroy forms of religiosity incompatible with its own moral and spiritual perceptions' (Taylor, 2010: p. ix). In this sense, *Crow* is not just a collection of poems *about* a dark green religious worldview, but rather an enactment of a dark green religious mythology, using the trickster story to challenge and abolish a prevailing Christian world order.

Dark green religion has many affinities with deep ecology or ecocentrism, terms more often used in ecocritical discourse for referring to beliefs in nature's 'inherent worth'. Like dark green religion, Garrard points out that

> The notion of ecocentrism has proceeded from, and fed back into, related belief systems derived from Eastern religions, such as Taoism and Buddhism,

> from heterodox figures in Christianity such as St Francis of Assisi (1182–1226) and Pierre Teilhard de Chardin (1881–1955), and from modern reconstructions of American Indian, pre-Christian Wiccan, shamanistic and other 'primal' religions.
>
> (2012: 25)

The identification of St Francis of Assisi as an atypical Christian, relating to spiritual or animistic traditions, is also made by White, who describes St Francis as 'the greatest radical in Christian history since Christ', and as 'The greatest spiritual revolutionary in Western history'. White concludes his argument by noting that 'Francis tried to depose man from his monarchy over creation and set up a democracy of all God's creatures', and moreover that his 'view of nature and of man rested on a unique sort of pan-psychism of all things animate and inanimate'. On these grounds White proposes 'Francis as a patron saint for ecologists' (1967: 1206–7). However, White's argument for St Francis as an exception does not change the fact that he nevertheless belonged to the Christian faith. In comparison, Hughes's proposal of an alternative to a relationship between humans and nature based on Christian ideology is much more radical, giving up a Judaeo-Christian belief system altogether and replacing it with a spiritual ideology based on an entirely different set of basic principles.

The difference between a 'dark green' and a simply 'green' religion corresponds to a difference between 'shallow' environmentalism and deep ecology, also described by Garrard:

> whereas 'shallow' approaches take an instrumental approach to nature, arguing for preservation of natural resources only for the sake of humans, deep ecology demands recognition of *intrinsic value* in nature. It identifies the dualistic separation of humans from nature promoted by Western philosophy and culture as the origin of environmental crisis, and demands a return to a monistic, primal identification of humans and the ecosphere.
>
> (2012: 24)

With similar reasoning, Taylor refers to 'green religions' as religions that have undertaken revisions and reforms in response to contemporary environmental concerns, but without changing their basic beliefs, while for dark green religions the belief in the inherent value and sacredness of the natural world is the most fundamental principle.

In addition to beliefs in nature's intrinsic value, dark green religion emphasises the role of myth and ritual in formulating and mediating the relationship between man and nature. Taylor describes this spiritual approach as

> generally deep ecological, biocentric, or ecocentric, considering all species to be intrinsically valuable, that is, valuable apart from their usefulness to human beings. This value system is generally (1) based on a felt kinship with

> the rest of life, often derived from a Darwinian understanding that all forms of life have evolved from a common ancestor and are therefore related; (2) accompanied by feelings of humility and a corresponding critique of human moral superiority, often inspired or reinforced by a science-based cosmology that reveals how tiny human beings are in the universe; and (3) reinforced by metaphysics of interconnection and the idea of interdependence (mutual influence and reciprocal dependence) found in the sciences, especially in ecology and physics.
>
> (2010: 13)

All of these principles are present in *Crow*, negotiated and formulated in poems where Crow considers the evolution of himself in relation to his surroundings, critiques Christian arrogance, and considers and observes the natural world.

Though there is no clear cut definition that separates religion from spirituality, dark green religion is more spiritual than religious in the sense that it is personal, even private, rather than institutionalised, according to Taylor: 'In common parlance, *religion* is often used to refer to organized and institutional religious belief and practice, while *spirituality* is held to involve one's deepest moral values and most profound religious experiences.' Taylor adds that while 'the world's predominant religions [...] are generally concerned with transcending this world or obtaining divine rescue from it', spiritual enterprises are more about 'personal growth and gaining a proper understanding of one's place in the cosmos', and it is this difference that allows spiritual beliefs 'to be intertwined with environmentalist concern and action' (2010: 3).

This differentiation between religion and spirituality suggests that *Crow* criticises not just Christianity but all religions that are institutionalised belief-systems and that focus on ideas of transcendence. The idea of the personal spiritual experience is enacted in the character of Crow, whose developing sense of his own morality and relation to the world around him is communicated through the different poems. To the extent that Crow develops a consciousness over the course of the book it is as a result of his personal journey, not through the discovery of an alternative to Christianity formulated by someone else.

Taylor differentiates between four types of dark green religion. Of these, *Crow* is best described as an expression of animism, defined as 'the perception that spiritual intelligences *or* lifeforces animate natural objects or living things', and specifically in line with Taylor's definition of a 'Naturalistic Animism'. Different from 'Spiritual Animism', naturalistic animism does not include any sort of supernatural realm, and often includes a Darwinian view of the history of the earth:

> Naturalistic Animism involves either skepticism or disbelief that some spiritual world runs parallel to the earth and animates nonhuman natural entities or earth herself. But those engaged in it nevertheless express, at minimum, kinship with and ethical concern for nonhuman life. Moreover, for many naturalistic animists, understanding and even communicating with nonhuman lifeforces is possible. According to the historian Donald Worster,

> this kind of felt kinship, and the biocentric ethics that tends to accompany it, can be grounded in evolutionary theory. [...] Darwin clearly believed that a kinship ethic can be deduced from knowledge of our common ancestor and awareness that other animals suffer and face challenges, as do we. This kind of conjecture represents an emphatic form of analogical reasoning as well as an act of moral imagination – this is typical of those engaged in Naturalistic Animism. Animism understood in this way can be entirely independent of metaphysical speculation or supernaturalistic assumptions.
>
> (Taylor, 2010: 15, 22–3)

The move away from Christian guilt and moral anguish portrayed in 'Crow Tyrannosaurus' and 'Crow's Nerve Fails' to Crow's view of himself as part of a larger context ordered according to evolutionary principles in 'Crow Frowns' reflects this move from a Christian to an animistic worldview, where everything that is sacred exists on earth.

Crow connects this naturalistic animistic worldview to the trickster narrative. Taylor suggests that animistic beliefs also entail that 'people can, *at least by conjecture and imagination*, and sometimes through ritualized action and other practices, come to some sort of understanding of these living forces and intelligences in nature and develop mutually respectful and beneficial relationships with them' (Taylor's emphasis). Through the trickster figure and songs (poems) of Crow, Hughes demonstrates a creative and aesthetic way to engage in the process of understanding such lifeforces, enacting dark green religion 'understood as a quest to deepen, renew, or tap into the most profound insights of traditional religions' (Taylor, 2010: 15–16, 3). Understanding modern concerns for the environment through ancient stories and beliefs was crucial for Hughes; he was convinced that 'The old method is the only one', referring to how ancient myths and rituals were used to 'tap into' and channel 'the elemental power circuit of the universe' (quoted from Faas, 1980: 200).

The naturalistic animism in *Crow* is most clearly expressed in the last poem, 'Littleblood' (Hughes, 2003: 258). In this poem the speaker attempts to communicate with the spirit of nature, what Taylor calls a 'lifeforce'. This nature spirit, referred to as 'littleblood', is present everywhere in the natural world, though it is currently hiding and wounded:

> O littleblood, hiding from the mountains in the mountains
> Wounded by stars and leaking shadow
> Eating the medical earth.

These lines capture the difference between littleblood and Christian saints, as described by White:

> It is often said that for animism the Church substituted the cult of saints. True; but the cult of saints is functionally quite different from animism. The saint is not *in* natural objects; he may have special shrines, but his citizenship

> is in heaven. Moreover, a saint is entirely a man; he can be approached in human terms. In addition to saints, Christianity of course also had angels and demons inherited from Judaism and perhaps, at one remove, from Zoroastrianism. But these were all as mobile as the saints themselves. The spirits *in* natural objects, which formerly had protected nature from man, evaporated. Man's effective monopoly on spirit in this world was confirmed, and the old inhibitions to the exploitation of nature crumbled.
>
> (1967: 1205)

Different from the mobile saints, littleblood is confined to earth; it has nowhere to hide from the mountains other than 'in the mountains'. While it has been injured by the stars, pointing to a Christian heaven, its only medicine is the earth.

The next lines explain that littleblood has no body of its own, 'O littleblood, little boneless little skinless', but is present everywhere:

> O littleblood, drumming in a cow's skull
> Dancing with a gnat's feet
> With an elephant's nose with a crocodile's tail.

The last lines describe littleblood as part of a nature that is both creative and destructive, where death and violence are necessary parts of life. In the essay 'Poetry and Violence' in *Winter Pollen*, Hughes reflects on positive and negative meanings of the word 'violence', noting that 'the word obviously covers a great range of different degrees of seriousness [...] in the way of moral and spiritual consequences' (1994: 253). At one extreme, there are acts of violence of the most serious and negative kind, such as war, while at the other end there are the least serious meanings, without negative implications, a 'violent rain' for example. At all its different stages of seriousness, Hughes suggests, the word can have either positive or negative meanings.

The kind of violence that Hughes portrays can be defined on this scale as serious but non-negative, or value-neutral. It is a violence that is equally necessary to create and destroy life, prompted by instincts evolved by natural selection, enacted, for example, by 'predators killing to eat' (1994: 256). For Hughes, the kind of violence that propels evolution is sacred because it shows 'the operation of divine law in the created things of the natural world' (1994: 259). In this sense the natural world, including its violent aspects, is sacred for Hughes. It is violence of this strong but value-neutral kind that is indicated in the last stanza of 'Littleblood' that juxtaposes destruction and creation.

The spirituality of this sacred, natural world is for Hughes intimately connected to the spirituality of poetry as a creative act, and it is the spirit of natural and divine creation that is beckoned to come and sing in the poet's ear in the last line of 'Littleblood': 'Sit on my finger, sing in my ear, O littleblood'. These last lines of *Crow* show a move away from the assertive tone and negative violence of poems like 'Revenge Fable', with 'Forbidding, screaming and condemning' humans, or 'Crow's Account of the Battle' where 'Theorems wrenched men in two', towards a quieter, more humble tone that is attentive to both inner and outer nature, trying to listen to the spirit of a sacred, natural world rather than to a civilised, educated Christian human voice.

Note

1 An earlier version of this chapter was published as 'Different Shades of Green: A Dark Green Counterculture in Ted Hughes's Crow', *Ecozon@: European Journal of Literature, Culture and Environment*, special issue on Green Counterculture, guest edited by Peter Mortensen, 4/1 (2013).

References

Barbour, I. (1973) *Western Man and Environmental Ethics*, Reading, MA: Addison-Wesley.
Brandes, R. (2011) 'The Anthropologist's Uses of Myth', in T. Gifford (ed.), *The Cambridge Companion to Ted Hughes*, pp. 67–80, New York: Cambridge University Press.
Carson, R. (1962) *Silent Spring*, Boston, MA: Houghton Mifflin.
Crutzen, P.J., and Stoermer, E.F. (2000) 'The "Anthropocene"', *Global Change Newsletter*, 41: 17–18.
Faas, E. (1980) *Ted Hughes: The Unaccommodated Universe*, Santa Barbara, CA: Black Sparrow Press.
Gamber, J.B. (2009) 'Trickster Founders of This New Earth: Excessive Nature in Vizenor's *Harold of Orange*', paper presented at the ASLE Conference 'Island Time: The Fate of Place in a Wired, Warming World', Victoria, BC, 3–6 June.
Garrard, G. (2012) *Ecocriticism*, 2nd edn, London and New York: Routledge.
Gifford, T. (2009) *Ted Hughes*, London and New York: Routledge.
Heise, U. (2014) 'Plasmatic Nature: Environmentalism and Animated Film', *Public Culture*, 26/2/73: 301–18.
Hughes, T. (1994) *Winter Pollen: Occasional Prose*, London: Faber & Faber.
Hughes, T. (2003) *Collected Poems*, London: Faber & Faber.
Kochan, Lionel (1997) *Beyond the Graven Image: A Jewish View*. Basingstoke: Macmillan.
Radin, P. (1956) *The Trickster: A Study in American Indian Mythology*, London: Taylor & Francis.
Ramsey, J. (1983) '*Crow*, or the Trickster Transformed', in K. Sagar (ed.), *The Achievement of Ted Hughes*, pp. 171–185, Athens, GA: University of Georgia Press.
Ruether, R.R. (1992) *Gaia and God: An Ecofeminist Theology of Earth Healing*, San Francisco, CA: Harper.
Rogers, J. (1973) 'Ecological Theology: The Search for an Appropriate Theological Model', reprinted from *Septuagesino Anno: Theologiche Opstellen Aangebsden Aan Prof. Dr. G. C. Berkower*, Kampen: J.H. Kok.
Sagar, K. (2009) *Ted Hughes and Nature: 'Terror and Exultation'*, Peterborough, England: Fastprint Publishing.
Sax, B. (2003) *Crow*, London: Reaktion Books.
Shaiko, R. (1987) 'Religion, Politics and Environmental Concern: A Powerful Mix of Passions', *Social Sciences Quarterly*, 68: 265–281.
Taylor, B. (2010) *Dark Green Religion: Nature Spirituality and the Planetary Future*, Berkeley, CA: University of California Press.
White, L. (1967) 'The Historical Roots of our Ecologic Crisis', *Science*, 155/3767: 1203–7.

2 Human history and environmental time

Postmodern nature in Heaney's bog poems

Heaney first began to explore Ireland's boglands in the late 1960s together with his friend the Irish painter T.P. Flanagan. Flanagan notes that 'Bored with the holiday brochure stereotype of the green Ireland', the bogs offered a 'visual surprise'. He also notes that a sense of the bog's 'ancient life' gave him a feeling of 'connectedness with a pre-Christian primeval past' (quoted from Parker, 1993: 87). In 1967, he dedicated his painting 'Boglands' to Heaney, who returned the gesture two years later by dedicating the poem 'Bogland' to Flanagan.

Flanagan's painting 'Boglands' is dominated by dull colours, primarily black, grey, dark green and brown, and evokes a sense of something both mysterious and ominous. Circular brush strokes suggest wells or whirlpools that you might fall or be drawn into. This depiction of the bog as both dangerous and alluring recalls images of the bog from Heaney's childhood, and also provides the basis for the imagery of the poem 'Bogland', as Heaney has explained:

> What generated the poem about memory was something lying beneath the very floor of memory, something I only connected with the poem months after it was written, which was a warning that older people would give us about going into the bog. They were afraid we might fall into the pools in the old workings so they put it about (and we believed them) that *there was no bottom* in the bog-holes.
>
> (1980: 56)

Already in this early poem, Heaney connects the bogs to a sense of both personal and collective histories and memories. As Flanagan's comment indicates, they also connect the bog to pagan cultures that predate the Judaeo-Christian religions that mark the troubled contemporary history of Northern Ireland. Like Hughes's *Crow*, Heaney's bog poems address the relationship between religious beliefs and the natural environment, and trace the roots of a particular contemporary political and cultural conflict surrounding that relationship to an ancient, primeval past.

'Bogland' is the closing poem of the collection *Door into the Dark*, published in 1969, and marks the onset of a fascination with the image of the bog that lasted for the rest of Heaney's career. He returned to it again in the next collection, *Wintering*

Out (1972), before it culminated in the series of bog poems placed at the centre of *North* (1975), where images and archaeological finds in Danish bogs are a way for Heaney to respond to the violence of the Troubles unfolding in Northern Ireland at the time. It is picked up again in several individual poems in later collections.

In the bog poems, the history and ecology of the bogs provide symbols, metaphors and parallels, and in some sense even explanations, for the contemporary political violence in Northern Ireland. Archaeological findings in the bogs, especially the so-called bog people, lead Heaney to identify what he calls 'an archetypal pattern' tying together 'the long rites of Irish political and religious struggles' (1980: 57–8). This archetypal pattern is based on recurring images of religious rites and sacrifices made in the name of the land or earth – several bog bodies discovered in Ireland and Denmark are believed to have been sacrificed as part of pagan fertility rites. In his poems, Heaney connects these ancient practices to contemporary events in Northern Ireland.

For Heaney, the bogs' ability to preserve nearly intact bodies and artefacts for thousands of years make them veritable reserves for history and memory. The stable conditions in the depths of the bog, where time appears to stand still, counteract what at first sight and in temporal comparison seem to be short-lived cultural histories played out on the surface. By representing a different, ecological time scale, the bogs provide a perspective that goes beyond the point of view of the single individual, generation or even society or culture. Offering relics from as far back as the Iron Ages, the bogs described in *North* are wells of historical significance that present a nearly timeless perspective on the relationship between culture, religion and the natural environment.

This chapter examines three particular ideas that animate Heaney's bog poems. First, the bogs' ability to preserve remnants from the past challenges our perception of time and unsettles the linear relationship we conventionally perceive between past, present and future. Second, the human bodies and artefacts from thousands of years ago that come to the surface from the depths of a relatively natural landscape disconcert our definitions of nature and culture and how landscapes evolve through time. Third, while the bog poems consider the relationship between human violence and culture they simultaneously reflect on their own ability to address those concerns, questioning the relationship between aesthetic representation and reality.

In this chapter, these three themes, relating to time, the nature–culture divide and the relationship between art and the outside world, are highlighted and analysed through a postmodern ecocritical approach. The analysis draws on Serpil Oppermann's proposal of a postmodern ecocriticism, presented in the essay 'Theorizing Ecocriticism: Toward a Postmodern Ecocritical Practice' (2006), as well as Thomas Docherty's description of 'The Grauballe Man' as a 'postmodern event' in 'Ana-; or Postmodernism, Landscape, Seamus Heaney' (1997). Human geographer Dianne Meredith discusses the relationship between the physical landscape of the bog and its representation in Heaney's poems. Her analysis in the essay 'Landscape or Mindscape? Seamus Heaney's Bogs' (1999) provides another important stepping-stone for the analysis in this chapter.

Most of Heaney's bog poems appeared as a series of poems placed in the middle of *North*, describing in detail several well-preserved ancient bodies recovered from bogs: 'Come to the Bower', 'Bog Queen', 'The Grauballe Man', 'Punishment', 'Strange Fruit' and 'Kinship' (Heaney, 1975: 31–45). The analyses in this chapter focus mainly on 'The Grauballe Man' and 'Kinship', in addition to two earlier poems by Heaney that also address the bogs: 'Bogland' (Heaney, 1969: 41–2) from *Door into the Dark* and 'The Tollund Man' (Heaney, 1972: 47–8), from *Wintering Out*. However, the connection between the themes and ideas of the images discussed and a postmodern ecocritical approach refer to the bog poems as a group.

The bogs

To understand the power of the image of the bog for Heaney, it is important to understand what makes bogs such special formations. Bogs began to form in Ireland about 6,000 years ago as a result of a combination of deforestation and climate change: a general increase in rainfall coincided with people beginning to clear woodland in order to make room for agricultural fields. In areas that have plenty of rain and modest temperatures (such as Ireland), paired with acidic and low-nutrient soil conditions, the rate of plant decomposition can become lower than that of plant growth, resulting in an accumulation of peat, or dead plant material. The layer of peat can sometimes be up to 12 metres deep, hidden under the cover of a thin surface layer of living plants. The peat consists of up to 98 per cent water, and in Irish bogs most of that water is held in dead fragments of sphagnum moss, also referred to as 'bog builder'. The accumulation of peat into such deep layers can take thousands of years, but once developed the conditions in the peat, protected by the thin surface layer, are very stable.

In Europe, bogs are considered threatened environments, and many European boglands have disappeared in recent centuries as a result of drainage and cutting of peat to be used as an energy source. Peat was the first widely used fossil fuel, depended on by the Dutch empire: John MacNeill has shown how in the seventeenth century peat 'helped underpin the geopolitical assertion of the Dutch on the world stage' (2005: 193). According to the Irish Peatland Conservation Council (IPCC), peatland currently covers about 5–8 per cent of the world's surface. In Europe, while they are still extensive in the Nordic countries, in other areas peatlands have been lost or decimated: 'Much of the European peat resource has vanished as technology and development have advanced. All natural peatlands in the Netherlands have been lost, Switzerland and Germany each have only 500 ha remaining. The UK has seen a 90% loss of blanket bog and a 98% loss of raised bog. Ireland has only 18% of its original peatland area left.' The IPCC suggests that Ireland contains some of the most important remaining bog habitats in Europe, but while conservation campaigns have been successful in some areas, pressure on the bogs 'continues unabated and without public knowledge'.[1]

In addition to the up to 350 individual species of moss referred to collectively as Sphagnum moss, the bogs support an extensive number of plants and animals specially adapted to the acidic, nutrient-poor environment in and on the bogs. In a single bog pool one scientist counted 32,000 different microscopic animals. In addition to hosting various kinds of moss, the bogs form specific habitats for many kinds of liverworts and lichens. As carbon sinks, the bogs could also be important as a counterforce to rising levels of carbon dioxide in the atmosphere, though the details remain uncertain. With regard to the bogs' historical and cultural significance, the IPCC notes that, in addition to being valuable natural environments, the bogs can 'expand our understanding of people, culture, economy and climate far back into prehistory'.[2]

The IPCC's appreciative descriptions of the bogs contradict competing ideas about bogs as dreary and negative, less beautiful than lush, green landscapes, and informing derogatory expressions such as 'bog Irish', or metaphorical descriptions such as 'bogged down'. In this respect, the word 'bog' is very different to the more recent term 'wetland', which has more positive connotations of environmental value and biodiversity. Heaney's early bog poems, as well as Flanagan's paintings, signify attempts to change this view, and highlight the bogs' peculiarity and value as unique and unusual natural environments, a view also proposed by Henry David Thoreau in 1862:

> When I would recreate myself, I seek the darkest woods the thickest and most indeterminable and, to the citizen, most dismal, swamp. I enter a swamp as a sacred place, – a sanctum sanctorum. There is the strength, the marrow, of Nature.
>
> (Quoted from Giblett, 1996: p. v)

Rod Giblett quotes this statement by Thoreau in a book titled *Postmodern Wetlands: Culture, History, Ecology*, an interdisciplinary study that argues for better protection of bogs and other wetlands. Giblett acknowledges Heaney's contribution to changing the popular image of the bog, noting that 'From the earliest of his published poetry Seamus Heaney has celebrated the centrality of the bog to the Irish landscape, to Irish history and to Irish cultural identity' (1996: 245). As Giblett points out, Heaney's poems provide new associations for the word 'bog', relating to something that is both environmentally and culturally valuable, and helping to reclaim for it the more positive connotations of 'wetland' rather than the less favourable associations of 'swamp'.

Landscape into mindscape

Heaney's first poem about bogs, 'Bogland' (Heaney, 1969: 41–2), appears in his second volume of poetry, *Door into the Dark*, from 1969. It equates bog with Ireland:

> Our unfenced country
> Is bog that keeps crusting
> Between the sights of the sun.

It describes various artefacts that have been retrieved from the peat, noting the bog's astonishing ability to preserve elements and evidence of history:

> They've taken the skeleton
> Of the Great Irish Elk
> Out of the peat, set it up
> An astounding crate full of air.
>
> Butter sunk under
> More than a hundred years
> Was recovered salty and white.

The poem balances cultural and natural history by placing human culinary objects (the butter) next to animal relics (the elk) and aligning descriptions of archaeologists digging 'inwards and downwards' to retrieve 'waterlogged trunks / of great firs' with the uncovering of multiple layers that each seem 'camped on before'. This intertwining of natural and cultural references creates an image of the bog as a place where human and non-human histories are inseparable.

In 'Landscape or Mindscape? Seamus Heaney's Bogs', human geographer Dianne Meredith examines Heaney's bog poems using an approach she describes as 'literary geography':

> Literary geography (as opposed to literary criticism) would examine these poetical descriptions of bogs in comparison to how closely the imagery fits the empirical reality of the landscape, offering alternative approaches to geographical understanding. While objective studies in geomorphology and ecological succession may help to explain the origin and development of bogs, literary geography considers place-creation to be also subjective, based on landscape perceptions which take into account humanistic responses as well as purely environmental factors. Landscape and territories can be appropriated as living entities, creating 'geographies of the mind' which are then reflected in the structuring of space.
>
> (1999: 126–7)

Compared to literary studies, Meredith places greater importance on biological and physical components of the landscape, but wants to examine the meeting point between physical components and literary depiction:

> In particular, literary geography considers environmental definitions to be subjective, based on a writer's relationship to the land as well as the land itself. Poetic license may stretch description of a regional landscape beyond the confines of objective reality, bringing to light a stronger objectivity, inclusive not only of the physical environment, but also of the social, psychological, and the historical climate. In this sense, 'false' or fictive

> geographies may in fact reveal central themes of the environment, beyond those articulated by purely scientific investigation.
>
> (1999: 127)

Meredith suggests that perceptions of a particular landscape shaped by an imaginative engagement can help to create a sense of 'territorial identity' that highlights meaning as 'obtained rather than intrinsic'. From this perspective, Meredith considers the relationship between the actual Irish boglands and the bogs in Heaney's poems, with the hypothesis that 'a comparison between the objective and subjective geography may reveal a stronger reality, structured somewhere in the borderland between the two' (1999: 131).

Meredith notes Flanagan's influence on Heaney's choice of the bog as a symbol for Ireland as an image that countered 'the romanticised tourist vision of an Ireland with rolling green hills, rippling brooks and lush vegetation' with something darker, more complex, and with a deeper sense of history. Starting with 'Bogland', Meredith notes how already in this poem the bog takes on a meaning that is almost fused with the image of Ireland, as descriptions of the butter retrieved from the peat transforms into a description of the bog itself as butter:

> The ground itself is kind, black butter
>
> Melting and opening underfoot,
> Missing its last definition
> By millions of years.

The sense of bottomlessness continues in the next lines that imagine water from the Atlantic Ocean actually seeping up through the bog:

> The bogholes might be Atlantic seepage,
> The wet centre is bottomless.

The description of the bog as 'bottomless' is a memory of how the bog was described to Heaney as a child, as noted in the beginning of this chapter. Meredith also reads it as a reference to the bog as a repository of historical information:

> As Heaney contemplated the investigations of the bog as historical repositories, he concluded that they were bottomless. As one digs back into history – or down in the case of bogs – the past merges with the present so that each layer is stripped, it seems already camped on. Therefore, digging back into origins both uncovers and creates a sense of place.
>
> (1999: 130)

The description of the bog as 'bottomless' is thus both a personal memory and a reference to the bog's cultural and historical significance.

Meredith remarks that, despite the reference in 'Bogland' to the bog as 'Missing its last definition / by millions of years', the bogs are actually not very old. Nor are they reminiscent of a time before humans began to change the landscape:

> Today the bogs are perceived as one of the last Irish wilderness areas but in fact, when humans first colonised Ireland, there was very little bog. The first farmers cleared woodland, not bogland. This woodland clearance is believed to be one of the chief causes for the development of bogs, along with the change in climate from drier to wetter conditions. As the bogs expanded, farming was forced to retreat. The bog had free rein to become wild, uninhabitable land.
>
> (1999: 132)

This is the main discrepancy that Meredith finds between the actual bogs and their portrayal in 'Bogland'. Granting Heaney his poetic freedom, Meredith concludes that 'Bogland' and the other bog poems transform the landscape of the bogs into a mindscape that 'provides Heaney with an effective metaphor to communicate a sense of a living landscape, continually in process, responding to topography, changing climate, and human influence' (1999: 133). This mindscape is further and more darkly developed in the later bog poems, starting with 'The Tollund Man' in *Wintering Out*, the collection immediately following *Door into the Dark* in which 'Bogland' appears.

'The Tollund Man'

In the acidic and anaerobic conditions in the peat, organic material can be preserved for long periods of time. Human bodies have been found with their clothes, facial features and even fingerprints and body hair intact after as much as 4,000 years. These human finds are commonly referred to as 'bog bodies' or 'bog people'. The first bog body portrayed by Heaney in a poem is the Tollund man, an unusually well preserved bog body recovered from bogs in an area in Denmark called Jutland. The body was found in 1950, and is believed to be from the fourth century BCE.

Heaney's poem 'The Tollund Man' (Heaney, 1972: 47–8) begins with the speaker imagining himself travelling to Denmark to see the Tollund man in the museum in the town where he is displayed:

> Some day I will go to Aarhus
> To see his peat-brown head,
> The mild pods of his eye-lids,
> His pointed skin cap.

Heaney had learned about bog bodies from the archaeologist P.V. Glob's book *The Bog People*, translated into English in 1969 and describing the retrieval, analysis and historical context of so-called bog bodies. Glob proposes that a significant number of the bog bodies were sacrificed as part of fertility rites. The

Tollund man is one such example, found with a noose around his neck. This detail is included in Heaney's poem, which describes the body as 'Naked except for / The cap, noose and girdle', and imagines the fate of the Tollund man as a sacrifice to an earth goddess:

Bridegroom to the goddess,

She tightened her torc on him
And opened her fen,
Those dark juices working
Him to a saint's kept body

These lines introduce a religious frame of reference that is not present in 'Bogland', emphasised in the second part of the poem through words such as 'blasphemy', 'consecrate', 'holy' and 'pray'. These words set the scene for the following lines, which are central in the poem as the speaker's attention shifts in both time and space, from Aarhus and the Tollund man to victims of sectarian violence in Ireland in the 1920s: the speaker prays that the ancient body of the Tollund man, sacrificed to ensure the fertility of the earth, would have the ability to make the deaths of the more recent victims in Ireland meaningful in some similar way:

I could risk blasphemy,
Consecrate the cauldron bog
Our holy ground and pray

Him to make germinate
The scattered, ambushed
Flesh of labourers,
Stockinged corpses
Laid out in the farmyards,

Tell-tale skin and teeth
Flecking the sleepers
Of four young brothers, trailed
For miles along the lines.

As Edna Longley has observed, these lines summon the Tollund man as 'scapegoat, privileged victim and ultimately Christ-surrogate', hoping to somehow connect his symbolic sacrifice to an earth goddess to more recent victims of Irish religious violence (1997: 42).

In the third part of the poem, the speaker's attention turns away from Ireland and back to Denmark, describing the Tollund man's 'sad freedom' on his last journey. In relation to the attempt to make the violence in the poem's second part meaningful in some sense, perhaps the idea of a 'sad freedom' is an attempt to suggest that to some extent the Tollund man was aware of his own death as meaningful in the sense that it would secure the fertility of his people's land, that it would make something 'germinate'. The speaker wishes for a similar kind of meaning to help him make sense of the violent deaths in Ireland.

Something of his sad freedom
As he rode the tumbril
Should come to me, driving,
Saying the names

Tollund, Grauballe, Nebelgard[3]

The poem concludes by again comparing ancient Denmark to contemporary Northern Ireland:

Out here in Jutland
In the old man-killing parishes
I will feel lost,
Unhappy and at home.

The descriptions of the bogs and the bog people in this and the other bog poems collect diverse references and contexts into a unified theme and image. This image culminates in *North*, Heaney's next volume of poetry and the collection that most directly responds to the events taking place in Northern Ireland. Heaney attributes the increasing complexity of the image of the bog in these poems to a commitment he had started to feel towards something he describes as deeper than the political. This feeling began to develop with the writing of 'The Tollund Man', Heaney states:

> When I wrote that poem I had a sense of crossing a line really, that my whole being was involved in the sense of – the root sense – of religion, being bonded to something. I felt it a vow; I felt my whole being caught in this. And that was a moment of commitment not in the political sense but in the deeper sense of your life, committing yourself to something. I think that brought me to a new possibility of seriousness in the poetic enterprise.
> (Quoted from Parker, 1993: 91)

This new territory and enterprise is further explored in *North*.

'The Grauballe Man' and 'Kinship'

The Grauballe man is another bog body from the same area in Denmark as the Tollund man, recovered in 1952. Also like the Tollund man, the Grauballe man is believed to have lived during the early Iron Age, and, discovered with his throat cut, seems to have suffered a violent, unnatural death.

The speaker in 'The Grauballe Man' (Heaney, 1975: 35–6) immediately appears much closer to his subject than does the speaker in 'The Tollund Man', perhaps reflecting the sense of clarity and confidence that Heaney now felt in this image – it is no longer a question of introducing the poem with a reference to 'some day I will', as in 'The Tollund Man'. The speaker in this poem jumps straight into a description of the body in front of him:

> As if he had been poured
> in tar, he lies
> on a pillow of turf
> and seems to weep
>
> the black river of himself.
> The grain of his wrists
> is like bog oak,
> the ball of his heel
>
> like a basalt egg.
> His instep has shrunk
> cold as a swan's foot
> or a wet swamp root.
>
> His hips are the ridge
> and purse of a mussel,
> his spine an eel arrested
> under a glisten of mud.

Thus far, the body is described in terms of natural and passive similes: 'His instep' is like 'a swans foot / or a wet swamp root'. In the fifth stanza, however, something else happens, as the body's 'head lifts', the chin suddenly 'a visor'. The word 'visor' is both reminiscent of the Tollund man's 'cap' and a suggestion of armour. Especially in the latter sense it signifies a break from the natural similes in the previous stanzas, introducing a new set of associations, amplified in the next line referring to the man's 'slashed throat'. Following the plaintive, musing descriptions of the body as almost a part of nature, the sudden intrusion of the unexpected adjective 'slashed' contradicts those previous descriptions, inscribing the body instead in a violent, human context.

Once this connection to human violence has been made the speaker retreats and returns to the natural similes, noting, in the next stanza, that the 'slashed throat' has 'tanned and toughened' so that now

> The cured wound
> opens inwards to a dark
> elderberry place.

The next stanza suggests the vivid image that the bog body presents to the speaker, asking:

Who will say 'corpse'
to his vivid cast?
Who will say 'body'
to his opaque repose?

Longley observes that, while the body in 'The Tollund Man' is an 'urgent presence', the bog body in 'The Grauballe Man' has been processed by the imagination, so that 'what was hypothetical in "The Tollund Man" – the consecration of "the cauldron bog"', can be seen to have 'hardened into accepted doctrine' in *North*.

The idea of the bog body as a vivid image in the speaker's mind is developed further in the following two stanzas, which compare the bog body to an infant: 'his rusted hair' is 'unlikely / as a foetus's', emerging from the bog, he appears 'bruised like a forceps baby'. The description refers both to the painstaking process of retrieving the bog body from the peat and the violent background that informs Heaney's interest in it as an image in relation to the Troubles: Heaney has acknowledged that he was looking at this time for 'images and symbols adequate to our predicament', which he found in the image of the ancient bog people (Heaney, 1980: 56).

In the next stanza, the Grauballe man appears 'perfected' in the speaker's mind: 'now he lies / perfected in my memory'. In the next, and last, two stanzas, this image becomes a dual image comprising both 'beauty' and 'atrocity':

hung in the scales
with beauty and atrocity:
with the Dying Gaul
too strictly compassed

on his shield,
with the actual weight
of each hooded victim,
slashed and dumped.

The last three lines describe contemporary victims of the conflict in Northern Ireland. The way in which these contemporary events are introduced into the poem is similar to how related references are made in 'Punishment' (Heaney, 1975: 37–8), one of the other bog poems in *North*, where the speaker states that:

I who have stood dumb
when your betraying sisters,
cauled in tar,
wept by the railings,

who would connive
in civilised outrage
yet understand the exact
and tribal, intimate revenge.

Conor Cruise O'Brien comments on these lines that 'It is the word "exact" that hurts the most: Seamus Heaney has so greatly earned the right to use this word that to see him use it as he does here opens up a sort of chasm'. He adds that in the bog poems 'hope succumbs': 'I have read many pessimistic analyses of "Northern Ireland", but none that has the bleak conclusiveness of these poems' (1997: 26). They seem to say that this is what humans have always been like: cruel and callous, abiding to religions that demand brutal sacrifices in the name of the earth.

The 'Dying Gaul' referred to in the penultimate stanza of 'The Grauballe Man' is a Hellenistic sculpture of a dying Celtic warrior from the third century BCE, around the same time when the Grauballe man is believed to have lived. This reference to idealised and aesthetically represented violence is countered by the brutality of the description of contemporary victims as 'slashed and dumped' in the last stanza, and it is these two sides that are 'hung in the scales / with beauty and atrocity'. The word 'slashed' from the earlier description of the Grauballe man's throat is reused, underlining the alignment of the different contexts and victims.

In the essay 'Ana-; or Postmodernism, Landscape, Seamus Heaney', Thomas Docherty suggests that a postmodern reading of 'The Grauballe Man' brings out the poem's anachronistic elements, which counteract what he describes as 'the "punctuality" of the Modern, which is concerned to map two points in time as if they were two stable points in space'. Docherty argues that 'A philosophy of postmodernism will raise the stakes of the poem, disabling the conventional reading of it as a neo-Modernist exercise in myth-making and replacing the usual banal reading of its politics with something literally more compelling' (1997: 206). Rather than focusing directly on what 'The Grauballe Man' says about the events in Northern Ireland, Docherty's postmodern reading focuses on how the poem treats and destabilises concepts of time, including ideas regarding history, memory and the relationship between past and present.

Docherty points out that the bog in 'The Grauballe Man' not only represents but also *is* an actual 'repository of history and continuity across time'. Adding to the history that is actually present in the bog, Heaney associates the landscape with multifarious historical, geographical and political dimensions that, although in his poems they relate specifically to Northern Ireland, are not immediately present in the primary image of the bog. These added dimensions, according to Docherty, are informed by an idea of Ireland and Northern Ireland as 'critical spaces' in the sense that they relate to 'a "critical difference" called "the border" between North and South' (1997: 207). The idea of the geographical border as critical and unstable relates to the other similarly unstable and critical borders addressed by the bog poems between past and present, nature and culture, and reality and representation.

Docherty suggests that within the neo-Romantic and Modernist traditions with which Heaney is usually associated, '"imagination" forges a link between the Subject of consciousness and History as its Object'. Docherty questions the legitimacy of this link and argues that the neo-Romantic/Modernist subject has been rendered both aesthetically and politically unavailable to Heaney by a postmodern destabilisation of the relationship between the subject and its history, between the real and its representation:

> The 'ground' for [Heaney's] poetry, history itself in the Irish context, has disappeared, gone underground. As a result, a series of reversals takes place in 'The Grauballe Man': what seemed a tomb is a womb; what seemed a man gives a kind of birth while also being the baby itself; to dig is to discover not the past at all (history) but rather 'the presence of the past' (anamnesis). When Heaney wrote the poem, he was deeply aware of the presence of the past, not just in terms of his search for 'images and symbols adequate to our predicament', but, rather, in terms of the very historicity of the present, his present as a moment of flux, his spatial present as a moment bifurcated, divided, a moment when space has gone critical, differential, historical rather than antiquarian.
>
> (1997: 208–9)

It is from perhaps this premise that 'The Grauballe Man' asks:

> Who will say 'corpse'
> to his vivid cast?
> Who will say 'body'
> to his opaque repose?

Docherty suggests that what this stanza really asks is whether history is really dead and in the past, or whether it is alive and vivid in the present, and moreover that it is in fact 'the very posing of the question which opens up the text to a postmodernism' (1997: 209). Materialised in the physical appearance of the bog body, the question seems to be posed not only by the poet, but by the landscape itself; the unearthed corpse, several thousand years old but with hair, fingerprints and even facial features intact seems to cast doubt on whether the past is really in the past.

According to Docherty's postmodern perspective, the significant discovery made by Heaney in 'The Grauballe Man' is not of 'an archaeological remnant of the past' but of 'the flow and movement of history, history as "becoming" even as he writes'. This flow of history is not linear; it 'becomes' even as Heaney writes in the sense that the past intrudes on the present, as demonstrated by the discovery and unearthing of the bog body, and by its associations to current events. 'The Grauballe Man' 'delineates not the past but the presence of the past as a living present and the mutability of that present, its fluidity or flux' (Docherty, 1997: 210, 211). Heaney actualises the fluidity and flux of the present through the connection he creates between the bog people and the current conflict, but it is also actualised through the act of literally bringing the ancient bodies into the present by digging them out of the bog. These acts that complicate the relationship between the past and present lead Docherty to characterise 'The Grauballe Man' as a postmodern event, a description that also characterises the bog poems as a series.

The notion of the past as present informs what Docherty calls an 'anamnesis of history', a concept he uses to refer to history as an imaginative recreation of the past in the present, suggesting that such an 'actualisation of the virtual' takes place in 'The Grauballe Man' (1997: 212). The process of actualisation, of bringing the past into the present, takes place as images of the body as passive change into descriptions of the body as active. Following the refusal in the seventh stanza to refer to the body as a corpse, images of progression and life are introduced as the body is compared to a baby rather than, as previously, to oak and rock. At this stage of the poem, the bog body is 'perfected' in the speaker's memory. The 'perfection' of the image of the Grauballe man in the speaker's memory indicates the presence of the past in the present, a process also indicated by Meredith, who states that when digging down into the bog, 'the past merges with the present'. The completion of the imaginative process that transforms the Grauballe man from a passive historical artefact into an active presence prepares for the 'actual weight' of the contemporary victims that enters the poem in the last stanza.

Docherty notes that a similar process is described by Heaney in the poem 'Viking Dublin: Trial Pieces' (Heaney, 1975: 21–4), also in *North*, in which the speaker refers to himself as Hamlet, a 'skull-handler' and 'parabalist', only capable of

> coming to consciousness
> by jumping in graves
> dithering, blathering

These lines, in the same collection as the series of bog poems, capture the meta-reflection on the relationship between a violent reality on the one hand, and acts of representation and aestheticisation on the other.

The delineation of the past and present is the first and in this context most important of three postmodern elements that Docherty identifies in 'The Grauballe Man'. He describes the second as 'a kind of montage' or 'a kind of dialectical process', created by a 'collision' of images from 'Jutland and Ireland, the Iron Age and the IRA' (1997: 2011). It is seen most clearly in the last lines of the poem in which the image of the 'Dying Gaul' 'collides' with the image of 'the actual weight / of each hooded victim'. The Dying Gaul has been written about by a number of other poets, including Lord Byron in *Childe Harold's Pilgrimage*. The image is thus surrounded by several layers of aestheticisation – not only is it itself an aesthetic representation of violence, but the additional intertextual layers refer to a tradition of such representation so that the image when it occurs in Heaney's poem is removed by several layers of aestheticisation from the initial violent context that the image depicts. These additional references that are associated with the Dying Gaul add to the number and complexity of the colliding images that Docherty describes. The layering of historical and cultural references surrounding the Dying Gaul also relates to the archaeological layers of physical findings in the bogs, its different layers of history described in 'Bogland'.

The collision of images forms the basis for the issue of justice that Docherty argues 'dominates the latter half of the poem' (1997: 2011). The issue of justice is the third postmodern element that Docherty identifies in 'The Grauballe Man', and together with the other elements of the bog poems that he defines as specifically postmodern it forms the basic premise of the logic of the bog poems as a series: the delineation of past and present, a collision of images from Iron Age Denmark and twentieth-century Northern Ireland, and the question of justice and meaning in relation to acts of violence motivated by cultural and religious traditions.

The last and longest in the series of bog poems in *North* is 'Kinship' (Heaney, 1975: 40–5). This poem, like 'Bogland' but unlike 'The Tollund Man' and 'The Grauballe Man', is about Irish bogs. Different from 'The Grauballe Man', which begins by describing the bog body as part of nature, this poem begins by reversing that order, instead reading nature as culture. The speaker feels 'Kinned by hieroglyphic / Peat on a spreadfield'. He steps, physically as well as imaginatively, 'through origins', examining his 'memories of wilderness / On the kitchen mat', where he finds that the water in the bog both 'cheeps and lisps'.

The speaker in 'Kinship' eventually finds himself standing 'at the edge of centuries / Facing a goddess'. Michael Parker suggests that the primary subject of the bog poems is 'the Republican tradition, and how the cruelties inflicted upon "dark-bowered queen", Mother Ireland, Kathleen ni Houlihan, Shan van Vocht, have brutalised her sons, engendering a love of territory and ancestry that can carry them to appalling extremes' (1993: 134). The tension and contrast between appreciation of the landscape and the 'appalling extremes' that the attachment to the same land can lead to are reflected in 'Kinship' by a contrast between celebrations of the bog and descriptions of it as marked by suffering. In the poem's first part the speaker states that "I love this turf-face, / Its black incisions [...] I love the spring / Off the ground'. In the second part the bog is also soft and peaceful:

> *bog*
> meaning soft,
> the fall of windless rain,
> pupil of amber.

However, the next stanzas present a darker image:

> Insatiable bride.
> Sword-swallower,
> Casket, midden,
> Floe of history.
>
> Ground that will strip
> Its dark side,

Nesting ground,
Outback of my mind.

The change between a peaceful, soft bog and a dark, 'insatiable' 'bone vault' suggests the complexity of the image of the bog. In the fourth part of 'Kinship', the speaker confirms the strength of the image by stating that 'This centre holds', a response to Yeats's phrase in 'The Second Coming' that 'the centre cannot hold', a reference perhaps both to the durability and stable conditions in the physical bog and the perseverance and timelessness of the cultural conflicts it bears witness to, spreading from past to presence:

This centre holds
And spreads,
Sump and seedbed,

A bag of waters
And a melting grave.
[...]
This is the vowel of earth
dreaming its root
in flowers and snow,

mutation of weathers
and seasons,
a windfall composing
the floor it rots into.

These descriptions of natural processes are complemented by references in the third and fifth parts of the poem to the speaker's own memories of the bog, of turf-carts and wagons, introducing a contemporary relationship between humans and the bog.

While there are some references in the previous parts of the poem to a darker side of the bog, the darkest version of the history of the speaker's land is painted in the sixth and last part of 'Kinship'. The title of the poem appears ironic as the bog is described as 'an old crannog / piled by the fearful dead':

Our mother ground
is sour with the blood
of her faithful,

they lie gargling
in her sacred heart
as the legions stare
from the ramparts.

Looking into the bog, the speaker sees victims past and present:

> Read the inhumed faces
>
> Of casualty of victim;
> Report us fairly
> How we slaughter
> For the common good.

If 'The Grauballe Man' moves between the past and the present, 'Kinship' moves from celebration to horror, from '*bog* / meaning soft', like 'the fall of windless rain', to a bog 'where nothing will suffice' and where victims, past and present, are sacrificed to a goddess who 'swallows', indiscriminately, 'our love and terror'.

A postmodern nature

Digging into the bogs, Heaney finds no remedy for current events. Instead he finds evidence of the historical and geographical persistence of what the poems suggest are similar behaviours, framed by religious contexts. The only bog poem that presents a landscape that is not characterised by human violence is the first poem, 'Bogland', and read in hindsight, it reads like a preparation for the deeper engagement with the image of the bog that followed in *North*. In the later bog poems, appreciation of what is first perceived to be a natural landscape repeatedly turns into realisation and recording of the 'appalling extremes' that have been carried out in the name of that same land. This shifting focus from descriptions of nature to recognition of a cultural presence is seen both in 'The Grauballe Man', where the carefully formulated natural similes in the first stanzas are vitiated by the sudden intrusion of the brute and careless 'slashed' in the fifth stanza, and in 'Kinship', where the bog that is initially perceived as soft and peaceful becomes a record of human sacrifice.

By showing that what is initially perceived to be non-human nature is actually permeated by a long history of human presence, the bog poems undermine the perception of nature and culture as separate realms. In 'Theorizing Ecocriticism: Toward a Postmodern Ecocritical Practice', Serpil Oppermann aims to rethink 'the ecocritical field of inquiry with the purpose of situating it within the broader perspective of a reconstructive postmodern theory', and 'conflate ecocriticism with an ecocentric postmodern theory for the purpose of developing a postmodern ecocritical approach which will help expand the ecocritical practice beyond its present limits'. Oppermann argues that 'postmodernism emphatically dismantles disjunctive opposites' so that 'nature is no longer perceived to be the Other'; by breaking down the barrier between conventional binary perceptions, postmodern analysis can open up a space 'for mutually constitutive relationships between culture and nature' (Oppermann, 2006: 104, 116).

Postmodern scholars have argued that ecocriticism needs to become more theoretical, especially with regard to the issue of representation, and would benefit from input from what they see as a more sophisticated postmodern tradition. Some traditional ecocritics, on the other hand, have argued that postmodernism undermines the relationship between the sign and its referent to the point of invalidating what Buell calls 'literature's referential dimension' (1995: 86). As a representative of the former, Oppermann argues that 'no interpretative theory can be conceived of without language occupying its center', and that 'It is precisely because of ecocritical underestimation of this fact that much work in this promising field of eco-literary studies does not go beyond simplistic contextual analyses of both literary and environmental texts' (2006: 103–4). This kind of analysis, Oppermann suggests, emphasises and encourages us to perceive and value nature as it appears to us, without changing it through our interpretation. It is the opposite process from that which Meredith describes as she traces Heaney's transformation of the bogscape into a mindscape. Indeed, the bog poems as a series present the opposite of the kind of realist epistemological view that Oppermann critiques, by undermining the relationship between nature's appearance and what it actually signifies and demonstrating that what seems to be non-human nature is really reserve of cultural histories and memories, counteracting the idea that nature simply 'is'.

To address the gap between 'poststructuralist emphasis on pure textuality and ecocritical focus on nature as a pure referent', Oppermann suggests a specifically postmodern ecocritical methodology characterised by 'a dialogic interaction of texts and contexts' and analyses that 'draw attention to the linguistic manipulations behind the discursive constitution of nature'. She suggests that this methodology would be especially relevant for texts that may not appear to have the natural world as their primary subject, but in which nature nevertheless plays a crucial role 'as a literary device' that affects the process of signification (Oppermann, 2006: 118, 119). In the bog poems, the discursive relationship between text and context that Oppermann describes as postmodern is complicated by the fact that the positions of text and context are in a sense reversed, as the natural landscape to some extent functions as text rather than context, as the speaker reads his own memories from the actual 'hieroglyphic / peat'. The landscape interprets his cultural history and contemporary setting as much as the other way around, by bringing the physical artefacts of the bog bodies to the surface as a kind of message that shapes the speaker's understanding of his current context. At the same time, the bogs also function as context, and it is through a dialogue with that context that the poems take form. The relationship between the bogs and what they signify is even further complicated by the fact that their message is already physically present in (though not limited to) the depths of the bogs; it is a both material and historical message that the speaker is not responsible for. Through these multiple layers of reversals and mutual discursive constitution the bog poems demonstrate a depiction of text and context that seems to exemplify what Oppermann describes.

Oppermann adds that at the bottom of 'the discursive constitution of nature [...] lies human oppression of the nonhuman world resulting in the environmental degradation' (2006: 117). The bog poems suggest a reversal of roles in this sense as well. Though the bog poems are not explicitly concerned with environmental degradation and do not directly express concern for boglands degrading due to drainage and peat extraction, by creating a 'mindscape' of the bogs the poems nevertheless contribute to a debate about valued landscapes. Moreover, the references to human sacrifices to an earth goddess and to contemporary victims also related to the land in some sense reverse contemporary issues of environmental oppression and domination – in the case of both the ancient pagan rituals and the contemporary context, humans are sacrificed for the sake of the earth rather than the other way around. Heaney observes this as a recurring pattern in a comment about Glob's book *The Bog People*:

> P.V. Glob argues convincingly that a number of [the bog people], and in particular the Tollund man, whose head is now preserved near Aarhus in the museum Silkeburg, were ritual sacrifices to the Mother Goddess, the goddess of the ground who needed new bridegrooms each winter to bed with her in her sacred place, in the bog, to ensure the renewal and fertility of the territory in the spring. Taken in relation to the tradition of Irish political martyrdom for that cause whose icon is Kathleen Ni Houlihan, this is more than an archaic barbarous rite: it is an archetypal pattern.
>
> (Heaney, 1980: 57)

This archetypal pattern is the main theme of the series of bog poems in *North*, reversing the pattern of human oppression and domination of nature that Oppermann describes, complicating our perceptions of the relationship between human and non-human agency.

Neil Corcoran has described the repeated replacement of nature by culture in the bog poems as creating a sense of claustrophobia, a sense further emphasised by the additional reversals outlined here. Corcoran suggests that the bog poems' discovery of 'an interconnectedness between contemporary sectarian atrocity in the North of Ireland, the behaviour of Viking invaders, and the ritual murders of Iron Age Jutland' creates a feeling of walking in circles, leading to an experience of the bog poems as 'claustrophobically obsessive and intimate' (1998: 62). A recurring pattern is seen not only in the behaviour to which the poems refer, but also in the structure of the poems themselves, as each poem begins with positive references to nature yet arrives at a place defined by human violence. This repetitive pattern creates a self-referential context that contributes to the claustrophobic feeling described by Corcoran, who suggests that 'The effect of this risky procedure is to make a certain monotony part of the poems' effect; they seem introverted, almost incestuously self-generated' (1998: 25). Through this introversion and repeated reversals of the relationships between human and non-human nature, the bog poems undermine any attempts to determine or rely on any kind of stable realist epistemology of the kind that Oppermann criticises. At the same time, however, by depending so extensively and explicitly on the physical presence and characteristics of the bogs, manifested especially by the detailed presence of the bog people and other archaeological findings retrieved from the peat, the bog poems resist any postmodern marginalisation or invalidation of the natural referent. Through this simultaneity and complexity the ideas and themes in the imagery of the bog poems exemplify one way to bridge postmodern and ecocritical readings.

Notes

1 Information from the Irish Peatland Conservation Council (IPCC), www.ipcc.ie/wpeurope.html; www.ipcc.ie/inforaisedbogfs.html [accessed Mar. 2012].
2 Ibid. www.ipcc.ie/infosphagnum.html; www.ipcc.ie/bogsimportant.html [accessed Mar. 2012].
3 All these are Danish place names.

References

Buell, L. (1995) *The Environmental Imagination: Thoreau, Nature Writing, and the Formation of American Culture*, Cambridge, MA: Harvard University Press.

Corcoran, N. (1998) *The Poetry of Seamus Heaney: A Critical Study*, London: Faber & Faber.

Docherty, T. (1997) 'Ana-; or Postmodernism, Landscape, Seamus Heaney', in M. Allen (ed.), *Seamus Heaney*, pp. 206–222, New York: St Martin's Press.

Giblett, R. (1996) *Postmodern Wetlands: Culture, History, Ecology*, Edinburgh: Edinburgh University Press.

Heaney, S. (1969) *Door into the Dark*, London: Faber & Faber.

Heaney, S. (1972) *Wintering Out*, London: Faber & Faber.

Heaney, S. (1975) *North*, London: Faber & Faber.

Heaney, S. (1980) *Preoccupations: Selected Prose, 1968–1978*, New York: Farrar, Straus & Giroux (first published by Faber & Faber).

Longley, E. (1997) '"Inner Emigré" or "Artful Voyeur"? Seamus Heaney's *North*', in M. Allen (ed.), *Seamus Heaney*, pp. 30–63, London: Macmillan.

MacNeill, J.R. (2005) 'Diamond in the Rough: Is there a Genuine Environmental Threat to Security? A Review Essay', *International Security*, 30/1: 178–95.

Meredith, D. (1999) 'Landscape or Mindscape? Seamus Heaney's Bogs', *Irish Geography*, 32/2: 126–34.

O'Brien, C.C. (1997) 'A Slow North-East Wind: Review of *North*', in M. Allen (ed.), *Seamus Heaney*, pp. 25–29, London: Macmillan.

Oppermann, S. (2006) 'Theorizing Ecocriticism: Toward a Postmodern Ecocritical Practice', *Interdisciplinary Studies in Literature and Environment*, 13/2: 103–28.

Parker, M. (1993) *Seamus Heaney: The Making of the Poet*, London: Macmillan.

3 Technology and landscape

Counter and recovery poems in *Elmet*

Elmet was the name of a Celtic kingdom in the north of England between the fifth and the early seventh century. The area is now part of West Yorkshire and includes the Upper Calder Valley on the south-eastern slopes of the Pennines, a low-rising mountain range stretching from northern England to the south of Scotland. The valley is named after the River Calder, which runs through it. One of the small towns along the banks of the river in the valley is Mytholmroyd, where Hughes was born.

The Upper Calder Valley was transformed by the industrial revolution in the late eighteenth and early nineteenth centuries. Major demographic changes took place as farmers abandoned the hills to take up work in mills and factories in the valley. By the latter half of the nineteenth century, the terraces surrounding the valley had been more or less abandoned, and cultivated fields reverted to moorland. Down in the valley, a railway and canal were built and roads were improved to facilitate transportation. Dams and other constructions were built to manipulate and secure the supply of water from the river. Later on, as mills were powered by steam rather than water, the burning of coal covered town buildings with a layer of soot.

With the decline of the textile industry in the first half of the twentieth century, the valley suffered. Again, demographic changes took place as people moved to nearby cities such as Leeds and Manchester. In Todmorden, one of the small towns in the valley, the population halved as mills were closed.

The history and landscape of the Calder valley is portrayed by Hughes in the collection *Remains of Elmet*, originally published in 1979 and followed in 1994 by a revised version simply titled *Elmet*. The 1979 edition was a collaboration with photographer Fay Godwin; Hughes wrote poems to accompany Godwin's black and white photographs of the Calder valley. While some of the poems recall the geological history of the valley, the volume focuses primarily on the transformations of the landscape that took place during and after the industrial era.

This chapter examines the relationship between landscape, technology and progress in these poems by Hughes. It suggests that the *Elmet* poems can be described as counter and recovery narratives, as identified by David Nye in relation to American frontier narratives. In *America as Second Creation*, Nye establishes a structural framework for identifying and characterising different narrative responses to technological developments. His theories focus on North America and refer primarily to industrialisation in the United States between the eighteenth and early twentieth centuries. This chapter finds Nye's theoretical framework to be illuminating also for the *Elmet* poems; the technologies involved are mostly the same (mills, canals and railways), as are concerns for their effect on local environments as well as on the workers. However, while Nye's primary examples are stories that formed contemporaneously with the technological developments that they describe, the *Elmet* poems were written after both the rise and fall of industrialisation in the Calder area, thus representing a retrospective point of view. This demonstrates Nye's points that the different types of stories he identifies are not necessarily progressive or chronological, but continue to be written and rewritten.

The foundation narrative

Nye outlines how, in response to technological progress in North America in and around the nineteenth century, different types of narratives emerged that described new inventions such as mills, canals, railways, dams and irrigation systems. The way these stories framed the new technologies differed significantly – some celebrated human ingenuity and progress, others depicted exploitation of workers, while yet others drew attention to environmental hazards.

The first type of story that Nye identifies is the foundation narrative, also referred to as master narrative or second-creation story. Foundation narratives provide new inventions with a cultural framework, a place or function in society. Nye describes the basic plot of a typical foundation story like this:

- A group (or an individual) enters an undeveloped region.
- They have one or more new technologies.
- Using the new technologies, they transform a part of the region.
- The new settlement prospers, and more settlers arrive.
- Land values increase, and some settlers become wealthy.
- The original landscape disappears and is replaced by a second creation largely shaped by the new technology.
- The process begins again as some members of the community depart for another undeveloped region.

(2004: 13)

This description corresponds to the process of industrialisation in the Calder valley. However, in the *Elmet* poems focus remains on one location, which means that they also describe what happens after the initial period of prosperity, when the frontier of development moves to a new region.

Generally, foundation stories embrace new technologies, and are characterised by a pioneering spirit and a will to conquer. They describe how new technologies found new communities and towns, often in areas that they refer to as former wilderness. Usually, they portray technology as improving both non-human nature and the quality of people's lives. According to Nye, such 'narratives naturalized the technological transformation of the United States so that it seemed an inevitable and harmonious process leading to a second-creation that was implicit in the structure of the world' (2004: 6). Foundation narratives, in summary, entertain a strong focus on positive developments that occur spontaneously and naturally and that benefit everyone involved.

Nye emphasises that this type of story was especially prevalent in America, where people to a further extent than in Europe viewed nature as something waiting to be improved by human designs. In the UK, things were different; Nye notes for example John Greenleaf Whittier's statement that 'When the rail-cars came thundering through his lake country, Wordsworth attempted to exorcise them by a sonnet'. In America people did not object in the same way to the formation of an industrial landscape. Nye suggests that the belief that nature was 'created not to be picturesque but to be useful' was more common in America than in the UK, supported by writers like Emerson, whom Nye suggests was among those who believed that 'the land exists for a purpose' (2004: 9, 10).

The type of narrative that this view of nature informed portrayed human exploitation of natural resources as a process that was in harmony with the order of the natural world:

> The persistent desire to assimilate nature to a second technological creation was the central feature of technological foundation stories. In each case, popular narratives explained how Americans were using new tools and machines to assimilate nature. These stories described the creation of new social worlds, ranging from frontier settlements to communities based on irrigation. In each case, a new form of society based on successful exploitation of a new technology became possible. The stories were central to the new nation's perception of history and geography, which is to say its perception of time and space.
>
> (Nye, 2004: 11)

The foundation story not only assumes but also promotes human dominion over nature. It argues that development equals progress, and sees progress as inherently positive.

The counter-narrative

The counter-narrative opposes the foundation story. It criticises the technology that the foundation narrative embraces, and brings to attention negative effects that a new invention may have on humans or on the environment. Rather than describing the prosperity that a new mill brings to a small town, for example, the counter-narrative might draw attention to poor working conditions inside the mill.

Nye defines counter-narratives as texts that 'resist or reimagine technological change and seek to ground identity not in machines but in other cultural artifacts and values'. Constructed by writers with different backgrounds and interests, including native populations, farmers, workers and environmentalists, counter-narratives depict conflicts rather than harmony between new technologies and existing natural and social orders. Especially in North America in the nineteenth century (but likewise in many other parts of the world at other times), 'second-creation stories treat the land as empty space, ignoring the original inhabitants', while counter-narratives, on the other hand, 'are told from the viewpoint of the indigenous community and/or emphasize the ecological effects of technological change' (Nye, 2004: 14). Compared to the outline of the foundation story, the counter-narrative may look something like this:

- Outsiders enter an existing biotic and/or human community.
- They acquire its land and assets by force or legal trickery.
- They possess powerful new technologies.
- They begin to use these technologies to transform the landscape, undermining the existing community's way of life.
- The existing community and the new one come into conflict.
- The new community wins.
- Additional settlers arrive and complete the transformation of the landscape.
- The original community loses population and goes into decline.
- Its people become marginal and disappear and move away.

(Nye, 2004: 15)

In the counter-narrative of the *Elmet* poems, the farmers on the slopes of the Calder valley represent the old, pre-industrialised community. The new community arrived with the businessmen who built the factories, mills and railway. Because the *Elmet* poems are written in retrospect, the last two points on Nye's list also apply to the community that was left after the mills closed and workers were forced to move elsewhere. This means that in the *Elmet* story it is not only the old community that eventually withers and disappears, but likewise the new community, with and in spite of its 'new powerful technologies'.

Further defining the counter-narrative, Nye suggests that it 'is often a tragic tale of struggle and defeat that begins with treaty violations or other illegalities', and then proceeds to 'reconstruct familiar events, sometimes by emphasizing a different ideological orientation or a fundamentally different epistemology' (2004: 15–16). Among the most pervasive such alternative ideologies are different forms of environmentalism, focusing both on the rights and values of non-human nature in and for itself, and in combination with environmental justice and human and social rights. One of the most famous environmental counter-narratives in recent history is Rachel Carson's *Silent Spring* from 1962, which includes both of these focal points. In the context of North American industrialisation, Nye suggests that counter-narratives

> saw new technologies not as tools used to make second creation but as weapons used to destroy the existing habitat. If Thoreau focused on the initial act of destroying the trees, and if Marsh focused on pollution, flooding, and other environmental effects, many later critics examined the rest of the mill narrative. How did the mill affect the society? Did it bring prosperity? Was the landscape of this second creation desirable? Most important, what were the effects on workers? These concerns gradually became predominant.
>
> (2004: 121)

Through different perspectives and emphases, counter-narratives identify aspects of technological development that corresponding foundation stories fail to recognise or acknowledge, or consider less important. These different stories create a dialogue between different frameworks and descriptions of new technologies.

The *Elmet* collections form a counter-narrative both as a single story, and through individual poems. The poem 'The Trance of Light' (Hughes, 2003: 459), for example, describes how the onset of industrialisation, rather than improving people's lives, put non-human nature to sleep:

> The upturned face of this land
> The mad singing in the hills
> The prophetic mouth of the rain
>
> [...] fell asleep
>
> Under migraine of headscarves and clatter
> Of clog-irons and looms

'The Trance of Light' suggests a deep ecological perspective on the relationship between man and nature as severed by industrialisation. Ralph Pite points out in an early British ecocritical study that in the view of deep ecologists 'the ecological crisis derives from the western understanding of man's place in the world', from the perception of a 'false division of man from nature'. In some contexts, it may be argued (as deep ecologists have done, as Pite notes (1996: 361), and as Hughes does in his collection *Crow*, discussed in Chapter 1) that this perceived division enables and legitimates the exploitation of nature as something separate from the human realm. By suggesting that industrialisation put the natural world to sleep, 'The Trance of Light' describes a similar view of a conflict between technological progress and the health of ecosystems to that expressed in the *Crow* poems, especially in 'Crow's Undersong', where 'nature' has been pushed to the peripheries of modern society.

Moving from an ecological to a sociological perspective, the *Elmet* poem 'Hill-Stone was Content' (Hughes, 2003: 463) describes a scene in which factories reduce workers to anonymous and interchangeable bodies ruled by machines, rather than the other way around:

And inside the mills mankind
With bodies that came and went
Stayed in position, fixed, like the stones
Trembling in the song of the looms.

Eventually people even begin to look like the mills: they become 'four-cornered, stony', standing immovable in 'the guerrilla patience / Of the soft hill-water'. Together with 'The Trance of Light', this poem describes a connection between outer and inner human nature, suggesting that, just as the process of industrialisation puts surrounding nature to sleep, it has a similar effect on the inner natures of people who work inside the factories, reducing them to parts of the machinery.

Nye illustrates the relationship between foundation and counter-narratives through the example of the axe. An improved axe inspired numerous North American settlement stories, typically describing a single male pioneer who ventures into the wilderness, clears the forest and builds a cabin. The cabin develops into an entire farm, and eventually into a new and prosperous town. Nye points out that this narrative provides the basis for the symbolic image of the log cabin, an American symbol of an idealised rural past. While this type of settler story portrays the natural environment as a resource waiting to be transformed into farmland, corresponding counter-narratives oppose that frame, and argue that the massive deforestation of North America in the nineteenth century signified thoughtless exploitation and recklessness rather than progress.

Stories about the mill provide another example. Different from stories about the axe that typically centred on one individual, Nye notes that 'the classic foundation narrative of the mill was widely accepted as a description of the formation and growth of communities' (2004: 97). According to Nye, foundation mill narratives justify the formation of towns 'as natural outgrowths of unexploited rivers and streams', often to the extent that the mills signify the completion of the world as created by God. As Nye points out, this story took a different form in America than in Britain:

> The antebellum mill narrative was the story of a partnership with nature. It was at once a fable of origins, explaining how towns and cities emerged from the landscape, as Americans developed its potentialities, and a counter-narrative to British developments. The foundation narrative gained force through a series of contrasts: not the 'artificial power' of steam but the 'natural power' of water, not vast factories but small mills, not smoky cities but pastoral towns, not an anonymous crowd of workers but women and children from a tightly knit community working for a short time in factory, not European landlords but American freeholders.
>
> (2004: 115)

The *Elmet* poems represent the opposite of the foundation narrative described here. In Hughes's view, mills represented an unnatural intrusion into an ancient landscape, fundamentally interrupting a pre-existing, harmonious relationship between human and non-human nature, reducing individuals to just workers, and polluting the natural environments both in the cities and on the surrounding moors. Nye notes that the British counter-narrative has an older history; he suggests that 'The primary meaning of this American foundational story lay in its inversion of the story of British industrial oppression' and, furthermore, that focusing on differences between England and America even 'distracted attention from the complaints of environmentalists and the protests of workers' (2004: 109, 116).

Hughes's poem 'Mill Ruins' (Hughes, 2003: 464) describes the decline of the textile industry in the Calder valley. Opposing foundation narratives that describe the rise of new communities, the counter-narrative in 'Mill Ruins' describes the space left behind after the mill has closed, when factories started to move to Asia in the twentieth century:

> One morning
> The shuttle's spirit failed to come back
> (Japan had trapped it
> In a reconstructed loom
> Cribbed from smiling fools in Todmorden).

The 'humming abbeys' that used to be busy with workers and products are described as 'tombs'. A sense of loss pervades the area; children trail 'aimlessly',

> Roaming for leftovers
>
> Smashed all that would smash
> What would not smash they burned
> What would not burn
>
> They levered loose and toppled down hillsides.

Nye suggests that 'If the foundation narrative presented the mill as the automatic source of prosperity and community development, the radical rewriting of that story presented it as the site of inevitable class conflict' (2004: 136). This story eventually made its way into numerous novels and other texts describing strikes and other forms of protests against working conditions in the mills. Nye concludes that the mill came to be seen more or less exclusively as a symbol of 'the willful exploitation of human labor': a source of poverty rather than prosperity, of cruelty rather than progression, and of pollution and corruption (2004: 128). As in *Elmet*, stories about mills transformed into counter-narratives.

While the counter-narrative in 'Mill Ruins' focuses on an emptiness left behind after the wave of industrialisation has passed, the short poem 'Wild Rock' (Hughes, 2003: 464–5) is concerned with what Nye refers to as 'the subordination of imagination to materiality' (2004: 176). In this poem the millstone is described as 'a soul-grinding sandstone'; working in the mill, the people are burdened by 'a permanent weight', their bodies as well as their minds merge with the stone building of the mill:

> A people fixed
> Staring at fleeces, blown like blown flames.
>
> A people converting their stony ideas
> To woollen weave, thick worsteds, dense fustians
>
> Between their bones and the four trembling quarters.

The same poem portrays environmental hazards, describing 'Grass greening on acid', and it is not just people who turn into factories and mills – even the sky above the valley has 'the face of a quarry'. The relationship between technologies, humans and the natural environment in 'Wild Rock' captures the depiction in the *Elmet* poems of how these different natures spill over into each other – working in a stony mill makes people's ideas similarly 'stony'.

'Remains of Elmet' and 'Top Withens'

'Remains of Elmet' (Hughes, 2003: 468–9), the title poem of the first edition, summarises the most significant events in the history of the Upper Calder Valley. Its creation by the end of the last Ice Age is referred to as the 'Death-struggle of the glacier', followed by cultivation by farmers, industrial transformation and post-industrial decline. If it had been a foundation narrative, the poem would have celebrated the foundation of towns and settlements in the previously uninhabited valley, affirming human progress and the conquest of land. Instead, it suggests that, despite all the efforts to the contrary, everything comes to nothing in the end, and only scraps remain of past endeavours:

> Now, coil behind coil,
> A wind-parched ache,
> An absence, famished and staring,
> Admits tourists
>
> To pick among the crumbling, loose molars
> And empty sockets.

While the absence in 'Mill Ruins' described the immediate absence of industry, the 'absence' in these lines refers to a void left by people, caused both by migration as a result of lack of work but also by the men from this area who had been killed in the two world wars (commenting on the effects of the wars on the valley, Hughes states in the introduction to *Elmet* that 'a single bad ten minutes in no man's land would wipe out a street or a village'; quoted from Gifford, 2009: 50). 'Remains of Elmet' communicates the opposite of growth or 'progress', through words such as 'death-struggle', 'corpse', 'cemeteries', 'vanished', 'nothing', 'sunk', 'crumbling' and 'empty'. No source of light or life seems to be indicated. Perhaps this lack of light or hope is the reason why 'Remains of Elmet' was not included in the later revised *Elmet*.

'Top Withens' (Hughes, 2003: 486–7) states explicitly that which is only suggested in 'Remains of Elmet', namely that 'it is all over'. The valley, once surrounded by 'hills full of savage promise', has arrived 'at the dead end of a wrong direction'. What used to be the 'dream's fort', despite being accompanied by 'dogged purpose', cannot keep the wind and the sky from 'swabbing the human shape from the freed stones'. The last line, however, signifies the difference from 'Remains of Elmet' – through the positive associations of the last adjective, 'freed', the poem suggests that, in spite of the hardship of human life in the valley, another form of life continues. The freed stones point forward rather than backward and suggest hope of ecological and geological renewal and regeneration.

Recovery narratives and anti-landscapes

Nye introduces the term 'anti-landscape' to describe certain types of damaged environments, defined as 'a man-modified space that once served as infrastructure for collective existence but that has ceased to do so, whether temporarily or long-term' (2010: 131). A typical example might be an abandoned mining site, which, polluted by toxic waste, can become more or less uninhabitable and include big, visible wounds in the landscape. Narratives that address such anti-landscapes are characterised by Nye as recovery narratives. Formulated in relation to damaged environments in general, and to counter-narratives in particular, Nye describes recovery narratives as 'Essentially about remaking a despoiled landscape', and therefore, unlike a foundation story, the recovery narrative 'begins not with empty space but with a place corrupted and degraded by human misuse'. A typical recovery narrative, according to Nye, may describe a site where

> the free market has unleashed selfish individuals who have exploited the land for short-term gain. To restore natural beauty and environmental harmony, a countervailing social force enters the area in the form of a nonprofit organization or government agency. After taking control away from short-sighted private interests, this institution redevelops the area. It cleans up pollution, halts erosion, plants new trees and shrubs, restocks rivers and lakes with native species, protects wildlife, and gives back to the public a restored version of the natural world.
>
> (2004: 294)

Nye mentions the American national parks Yosemite and Yellowstone as examples of places where such processes have taken place, thanks in part to writer and environmentalist John Muir.

Recovery narratives may not just record, but also inspire a process of social and ecological regeneration and restoration. Unlike foundation stories, the starting point for a recovery narrative is the recognition of a problem, environmental or other. Contrary to counter-narratives, the recovery story then aims to overcome, rather than just identify, that problem. As Nye points out, the recovery narrative dominated the twentieth century.

The most extreme anti-landscapes are poisoned, toxic spaces. The Upper Calder Valley of Hughes's *Elmet* poems is not uninhabitable, but it nevertheless exhibits some of the characteristics of an anti-landscape. As a result of its industrial history, it is affected by pollution, suggested by Hughes in references such as to 'the poisonous Calder' in the poem 'Sunstruck' (Hughes, 2003: 475–6), and as seen from buildings blackened by soot.[1] Many inhabitants also had to leave the valley in order to find work as jobs disappeared when mills and factories were closed. On grounds of both social and ecological conditions, the Upper Calder Valley may be described as a type of anti-landscape.

'Regeneration'

In the essay 'Regeneration in *Remains of Elmet*' (1994), Ann Skea states that

> Throughout his creative life, Ted Hughes has used his poetry to tap the universal energies and to channel their healing powers towards the sterility and the divisions which he sees in our world. All his major sequences of poetry work towards this end, and *Remains of Elmet* represents an important step in Hughes's ability to achieve wholeness and harmony through the imaginative, healing processes of his art.

According to Skea, *Remains of Elmet* illustrates the fate of a society that tries to dominate rather than cooperate with nature, and represents a 'demonstration of Nature's supremacy over humankind'. She suggests that 'Hughes's poems recreate [the world of Elmet] vividly in such a way that we may perceive the human errors which have desecrated it and the enduring, everpresent forces of Nature which survive' (1994: 116). The two main themes that Skea identifies in *Remains of Elmet*, the simultaneously destructive and creative powers of nature, correspond to the characterisation of the *Elmet* poems as counter and recovery narratives.

One of the poems that most clearly expresses the creative and regenerative powers of nature in the *Elmet* collections is the last poem in the first edition, 'The Angel' (Hughes, 2003: 492–3). As Skea points out, the angel signifies an earth goddess or creative spirit. It is also a kind of 'mother nature', thus referring back to the first poem in the same collection, 'Where the Mothers', providing a frame for the whole collection. 'The Angel' describes a dream of 'something disastrous':

> It was an angel made of smoking snow.
> Her long dress fluttered about her ankles,
> Her bare feet just cleared the moor beneath her
> Which glowed like the night-cloud over Sheffield.

The angel represents a spiritual dimension of the physical landscape, described as both terrifying and beautiful. Skea describes it as 'a symbol through which [Hughes] invokes the elemental forces of the universe to redress the natural balance which has been disturbed' (1994: 122). From having appeared in flames in the beginning of the poem, in the end, the angel/goddess disappears 'Under the moor'. Skea points out that

> there are resemblances between Hughes' angel and the Apocalyptic messengers of several major religious texts. They, too, appear amidst fiery disturbances to bring a warning to the human race. They, too, serve an omnipotent power which threatens death and destruction for human misdeeds. And they, too, offer the hope of blessing and spiritual rebirth. Through such links as these, Hughes channels the invoked energies of his angelic symbol towards creative rather than destructive ends. At the same time, he indicates the spiritual aspect of the regeneration he seeks to effect in *Remains of Elmet*.
>
> (1994: 119)

This interpretation suggests that 'The Angel' is both a counter and a recovery narrative: the 'warning to the human race' is a response to how the valley has been exploited without regard for the natural environment, while the notion of spiritual regeneration and the creative powers of nature suggests a recovery narrative.

Keith Sagar has characterised *Remains of Elmet* as being

> entirely about the crime against nature, which here takes the form of the enslavement of a people conscripted into the mills, the chapels, the trenches, conscripted also into the human attempt to conscript in turn the mothers, the sustaining elements of earth, air, fire and water, to degraded spiritless purposes.
>
> (2006: 153)

This crime committed against nature, according to Sagar, is the attempt to confine, contain and suppress 'the mothers', or the natural world. In the *Elmet* poems, any attempt to modify or control forces of nature has a correspondingly restricting effect on the imaginations of the people that inhabit the same environment, referring them, as Sagar puts it, to 'degraded spiritless purposes'.

Terry Gifford shifts the emphasis somewhat away from 'the crime against nature' when he states that the poems in *Remains of Elmet* 'celebrate the landscape and its elemental process as the dominant force to which human culture must adjust and thus be shaped by' (2009: 48). Borrowing the same expression from the *Elmet* poem 'Chinese History of Colden Water' (Hughes, 2003: 738–9) that also lends itself to the title of Sagar's book-length study, 'the laughter of foxes', Gifford concludes that 'Ultimately, Elmet is the kingdom of foxes and their laughter appears to be mocking the human presumption that it could be a home for industry, farming or religion' (2009: 51). While Sagar emphasises Hughes's depiction of environmental damage inflicted upon the landscape, seeing the *Elmet* poems as counter-narratives that highlight negative environmental consequences of technology and industry, in Gifford's view the *Elmet* poems are primarily about the ultimate temporality of that impact. This view sees the *Elmet* poems instead primarily as recovery narratives that counteract humans' attempts to control nature with much more powerful 'elemental forces' that sooner or later will overcome and repair any 'damage' humans are able to inflict.

Recycling: 'dead farms, dead leaves'

A different way of reading the *Elmet* poems as a recovery narrative is by seeing them through the perspective of recycling rather than regeneration. Throughout the collection, emphasis is put on recycling and organic renewal, on old factories falling into the earth. Gifford notes this and states that in these poems 'Hughes observes the temporary nature of the waves of human uses of the landscape', especially the insignificance of 'the road, railway and canal in the fullness of [the valley's] own geological timescale' (2009: 48). This relationship is reflected in 'Shackleton Hill' (originally titled 'Dead Farms, Dead Leaves', Hughes, 2003: 469–70). Gifford points out that this poem narrates the interconnectedness between the human and the natural world by intertwining social and ecological features of the landscape:

> Dead farms, dead leaves
> Cling to the long
> Branch of the world

Similarly, the earthly and infinitesimal is connected to the vast and astronomical:

> Stars sway the tree
> Whose roots
> Tighten on an atom

According to Gifford, 'Shackleton Hill' suggests that 'the failed human projects in this landscape can be understood as an organic part of the processes at work in this land'; farms and industries are simply different generations or branches on the same tree. While animals, people and factories 'Visit / And vanish', the tree continues to sway in the wind, directed by stars, providing a backdrop of continuity and stability. As Gifford suggests, 'Shackleton Hill' establishes a context in which sociological and ecological factors eventually become interchangeable and symptoms of the same causes, thus intermingling culture and nature. Based on this 'interchangeability of images', Gifford argues, 'dead farms *are* the dead leaves of the culture of Elmet', forming a linguistic pattern which persists throughout the *Elmet* poems (2009: 49, 136).

This interconnectedness between the human and the non-human also offers a way to understand 'Lumb Chimneys' (Hughes, 2003: 456–7), a poem that, as Gifford points out, 'provides an organic image for industrial decay' (2009: 49). In the poem, 'The huge labour of leaf is simply thrown away' and 'Great yesterdays are left lying', while, similarly described as passing incidentals, 'Brave dreams and their mortgaged walls are let rot in the rain'. Meanwhile, the 'spirit does what it can to save itself alone', and in the end, 'Nothing really cares. But soil deepens'. Despite the futility of human dreams and efforts, in the end they contribute, like falling leaves, to deepening the soil from which new life will grow. Having established this framework, the poem can focus on the context of industrial decay in the Calder valley and state, reassuringly, that 'Before these chimneys can flower again / They must fall into the only future, into earth'.

The processes of regeneration and recycling in the outer landscape are mirrored by changes in the inner lives of the people who live in the same area. As Gifford notes, and as already discussed, the poems suggest that, by trying to contain or suppress forces of nature, the human spirit is equally constrained:

> In Hughes's notion of Nature the outer processes echo the inner processes so that they are part of a whole. The organic tensions in external Nature are enacted in human nature. Processes at work in landscapes are at work in the human complex of energies. This is what 'The knight' recognises in *Cave Birds*, and Lumb in his identification with the tree from which his changeling self is made (*Gaudete*). In *Remains of Elmet* external natural processes are shown to be reflected not just in individuals but in both a farming and an industrial culture. Hence the poem title 'Dead Farms, Dead Leaves', for example, or the poem about 'Lumb Chimneys' 'flowering' like trees and 'falling into the future, into earth'.
>
> (2009: 136)

Hughes suggests that the ecological recovery of the industrialised valley depends on the imaginative reawakening of its local population. As Skea emphasises in her analysis, the *Elmet* poems are an attempt to inspire such an awakening.

The *Elmet* collections also claim the reverse, that spiritual reawakening is encouraged by the idea of ecological regeneration. The poem 'Long Screams' (Hughes, 2003: 460), for instance, begins by describing 'dead things':

> Unending bleeding.
> Deaths left over.
> The dead piled in cairns
> Over the dead.
> Everywhere dead things for monuments
> Of the dead.

However, by the end of the poem the mood is transformed by the creation of new life:

> The whole scene, like a mother,
> Lifts a cry
> Right to the source of it all.
>
> She has made a curlew

'Hardcastle Crags' and the voice of nature

The *Elmet* poems attempt two things. First, they illustrate people's exposure to forces of nature, through contrasting images of life on the valley floor and the harsh conditions on the surrounding moors. While changes (technological and other) take place in the valley, wind, rain and other natural elements prevail on the moors. From the perspective of the moors, people and their ambitions appear fleeting and even futile. In 'Wind' (originally published in *The Hawk in the Rain*, later included in *Elmet*, Hughes, 2003: 36–7), the natural elements suddenly reach down into the valley, interrupting people and reminding them of their vulnerability. The wind threatens to break into the houses:

> deep
> In chairs, in front of the great fire, we grip
> Our hearts and cannot entertain book, thought,
> Or each other. We watch the fire blazing,
> And feel the roots of the house move, but sit on,
> Seeing the window tremble to come in,
> Hearing the stones cry out under the horizons.

References to the history and ecology of the moors, and images of the powers and beauty of the non-human world, suggest the comparative smallness of human endeavours, counteracting any foundation narrative of industrialism or technological progress.

Second, the *Elmet* poems inscribe human nature and culture into the larger context of the natural environment, including the negative impacts they are sometimes responsible for, such as pollution and other forms of environmental degradation. This notion inspires recovery narratives in which the rise and fall of the textile industry becomes part of the wider ecosystem of the valley and the surrounding moors. According to this story, time and nature (referred to as the creative spirit, earth goddess or mother nature) will eventually heal the ecological wounds inflicted by industrial exploitation. As Gifford points out, in this way the *Elmet* poems in a way attempt to erase the line between nature and culture. Some research suggests that the Elmet region is shaped by human history to a further extent than previously appreciated; in a study of the history of the rural English landscape, David Hey notes that new discoveries of the influence of human activity on the production of open moorland during the Bronze Age 'has played an important part in the modern recognition of the huge scale of the prehistoric contribution to the development of the English rural landscape' (2002: 192). The relation between nature and culture in the *Elmet* poems could be developed in two directions, suggesting that cultural impacts may be viewed as part of more far-reaching natural processes, but also that the cultural heritage of this region is an integrated part of the history of the natural landscape.

'Hardcastle Crags' (Hughes, 2003: 456) also describes a recovery process. In this poem, polluting the environment is described as silencing the voice of nature. The poem begins by referring to a different type of silence, quoting a Taoist proverb: 'Think often of the silent valley, for the god lives there'. For Hughes, as noted by Leonard Scigaj, Taoism represents an antithesis to both Protestantism and mechanical science (1986: 235) (which also connects this poem to the depictions of religious and spiritual alternatives to Christianity and science and technology in *Crow*). The silence referred to in the opening line of 'Hardcastle Crags' is not really silence at all, but a silence that contains the voice of the natural world. 'Hardcastle Crags' suggests that this silence or voice of nature is now mixed up with a different sort of silence, left behind by the abandoned textile industry:

> here the leaf-loam silence
> Is old siftings of sewing machines and shuttles,
> And the silence of ant-warfare on pine-needles
> Is like the silence of clogs over cobbles

As factories are abandoned and the natural environment begins to recover, the silence of post-industrialisation gradually melts into the underlying silence of the valley: 'the beech-tree solemnities / Muffle much cordite'. The comparison of 'the silence of ant-warfare on pine-needles' to 'the silence of clogs over cobbles', which is not really silence at all, suggests that 'ant-warfare' is not silent or insignificant either, only that it is a sound on a different scale than human industrial sounds. The next stanza further emphasises that the silence of the valley is really a conglomeration of voices, past and present and from different representatives of the natural environment, that have been supressed by human noise:

> In a deep gorge under palaeolithic moorland
> Meditation of conifers, a hide-out of elation,
> Is a grave of echoes.
> Name-lists of cenotaphs tangle here to mystify
> The voice of the dilapidated river
> And picnickers who paddle in the fringes of fear.

In the final stanza, this natural world begins to recover as 'beech-roots repair a population / Of fox and badger'. There is also a sense of hope in the last lines, where 'a generation of slaves / Whose bones melted in Asia Minor' are transformed into a sound of 'love-murmurs' carried forward by the wind.

With a different focus, other *Elmet* poems celebrate the creation and continuity of life, human and other, in spite of what appears to be harsh conditions. In 'Football at Slack' (Hughes, 2003: 474–5), a football game happens up on the moors in spite of the fact that both players and ball are tossed about by the wind. In 'Wadsworth Moor' (Hughes, 2003: 474) the 'harebell and heather' signify 'a euphoria' in a landscape where 'the millstone of sky' has ground 'the skin off earth' so that the moor is

> A land naked now as a wound
> That the sun swabs and dabs
>
> Where the miles of agony are numbness

In 'The Big Animal of Rock' (Hughes, 2003: 466), 'root and leaf' are described as the 'offspring' of the rock, which in turn is 'singing' in its place among its 'ancestors'. In the poem 'Heather' (Hughes, 2003: 467–8),

> out of a mica sterility
> That nobody else wants
> Thickens a nectar
> Keen as adder venom

These descriptions change our perspective of the precedence of human agency in a landscape that is mainly passive, acting as backdrop to human societies. In these poems, even the inorganic objects are singing; they are portrayed as having a sense of agency that actively shapes their surroundings. In 'Heather', moreover, this agency is more far-reaching than the limited perspectives of humans:

> Heather is listening
> Past hikers, gunshots, picnickers
> For the star-drift
> Of the returning ice.

The heather is not interested or threatened by minor events such as wars or tourists, but worries instead about the next Ice Age.

These examples show how the *Elmet* poems attempt to redeem the Upper Calder Valley by focusing on the enduring and regenerative powers of the natural environment. Through this focus, Hughes is able to express a sense of faith in nature's ability to renew itself and repair human impact, while at the same time presenting a counter-narrative that opposes the characterisation of technological and industrial development as progress. However, the recovery narrative of the *Elmet* poems is a deep ecological recovery narrative, suggesting that while nature might recover, it will not necessarily do so in a way that includes human societies, and in this sense also suggesting a sense of warning.

Note

1 Many of these conditions have improved significantly since the writing of *Remains of Elmet* in the 1970s.

References

Gifford, T. (2009) *Ted Hughes*, London and New York: Routledge.
Hey, D. (2002) 'Moorlands', in J. Thirsk (ed.), *Rural England: An Illustrated History of the Landscape*, pp. 150–166, Oxford: Oxford University Press.
Hughes, T. (2003) *Collected Poems*, London: Faber & Faber.
Nye, D.E. (2004) *America as Second Creation: Technology and Narratives of New Beginnings*, Cambridge, MA: MIT Press.
Nye, D.E. (2010) *When the Lights Went Out: A History of Blackouts in America*, Cambridge, MA: MIT Press.
Pite, R. (1996) 'How Green were the Romantics?', *Studies in Romanticism*, 35/3: 357–73.
Sagar, K. (2006) *The Laughter of Foxes*, 2nd edn, Liverpool: Liverpool University Press.
Scigaj, L. (1986) *The Poetry of Ted Hughes: Form and Imagination*, Iowa City: University of Iowa Press.
Skea, A. (1994) 'Regeneration in *Remains of Elmet*', in K. Sagar (ed.), *The Challenge of Ted Hughes*, pp. 116–129, New York: St Martin's Press.

4 Colonised nature

Heaney and postcolonial ecocriticism

Reflecting on his collection *Field Work* from 1979, Heaney mentions a 'bay tree that grows half-symbolically in a couple of the poems' (Heaney, *The Poetry Book Society Bulletin*, 1979, quoted in Corcoran, 1987: 119). He is referring to the poem 'Elegy' (Heaney, 1979: 32):

> you found the child in me
> when you took farewell
> under the full bay tree
> by the gate in Glanmore

or maybe the 'Glanmore Sonnets' (Heaney, 1979: 41), where the tree also appears (in section IX):

> We have our burnished bay tree at the gate,
> Classical, hung with the reek of silage
> From the next farm, tart-leafed as inwit.
> Blood on a pitch-fork, blood on chaff and hay,
> Rats speared in the sweat and dust of threshing –
> What is my apology for poetry?

If this bay tree is only half-symbolical, it is also half-real. This chapter suggests that Heaney's use of the prefix 'half-' in his reference to this bay tree indicates a persisting theme in his ecopoetics – the intertwining of symbolical and actual ecologies. Combining perspectives from postcolonial and ecocritical theory, this chapter addresses the intermingling and intertwining of colonial and ecological themes in Heaney's poems from *Wintering Out* to his last collection, *Human Chain*, from 2010.

Postcolonialism and ecocriticism

The combination of postcolonial theory with ecocriticism was suggested already by Lawrence Buell in *The Environmental Imagination* from 1995, one of the foundational studies of ecocriticism. Buell notes that 'In my version of the history of the Western hemisphere, the ecological colonization of the Americas by disease and invasive plant forms is as crucial as the subjugation of their indigenous peoples by political and military means', and furthermore that 'William Bartram's botanical conquest of Florida is as notable an event of the American Revolutionary era as Patriot resistance to Britain' (1995: 6). More recently, several studies explore postcolonial ecocriticism in more detail: Graham Huggan and Helen Tiffin's *Postcolonial Ecocriticism: Literature, Animals, Environment*, and Roos and Hunt's *Postcolonial Green: Environmental Politics and World Narratives*, both from 2010, are just two examples. In the latter, Ursula Heise describes the overlaps between environmental concerns and social and justice issues that speaks to both these fields:

> Increasingly, ecocritics have come to emphasize that environmental problems cannot be solved without addressing issues of wealth and poverty, over-consumption, underdevelopment, and resource scarcity, while postcolonial critics have highlighted the ways in which historical struggles over colonial and neocolonial power structures as well as contemporary conflicts over economic globalization have involved and continue to revolve around fundamental environmental questions of, for example, land ownership, energy needs, uses of natural resources, agricultural production systems, pollution, exposure to risk, and local and global patterns of consumption. This intertwining of concerns over social justice and environmental conservation has led scholars to speak of 'a productive overlap' and 'opportunities for a fruitful alliance between two critical/theoretical schools'.
>
> (2010: 251)

In another article, Graham Huggan also notes the common interests between postcolonial and ecocritical studies, but highlights some of the differences that have divided the fields. The most important of these, Huggan suggests, is the ecocentric tendency of ecocritics, compared to an anthropocentric priority in postcolonialism. Where there is conflict between conservation and development, Huggan notes that 'ecocriticism has tended as a whole to prioritise extra-human concerns over the interests of disadvantaged human groups, while postcolonialism has been routinely, and at times unthinkingly, anthropocentric' (2009: 17). In Heaney's poems, the anthropocentric and the ecological are inextricably linked, and his poetics speaks to the emerging combination of these two theoretical fields. This chapter seeks to address a predominant critical focus on colonial themes in Heaney's poetry, which has sometimes been at the expense of recognising the important role ecological changes and conditions play in many of the poems.

Many critics have studied Heaney's poetics in relation to Northern Irish political and cultural history. Neil Corcoran states for instance that 'Heaney's painstakingly articulate self-consciousness about his "confidants and mentors" is impelled by the fact that, as a Northern Catholic, a poet who considers himself Irish, Heaney lies at an oblique angle to the English poetic tradition, and must labour to create his own personally sustaining "tradition" of sought-out exemplars' (1987: 120). Christopher Malone describes how ideas of nation, place and identity colour Irish poets' search for a cultural tradition and history:

> While Romantic Ireland is lost for Yeats, for poets who follow him the mythos of place continues to shape conceptions of national community. Northern Irish poets in particular have struggled in its absence to find authority to inscribe the space of home, responding to cultural pressures associated with Yeats's legacy. Certain questions recur for these poets: What does it mean to adhere to tradition in Ireland's radically changing present? Or to mythologize, when romantic conceptions of identity and place have serious consequences in modern Irish history?
>
> (2000: 1083)

He also adds that Heaney 'has come to assume Yeats's place as national poet and to typify, for many critics, a Northern Irish poetics' (2000: 1084).

In another study, Michael Allen suggests that the main theme of Heaney's poetry is how 'Britain and Ireland engage with their postcolonial legacy as the intercommunal conflicts endemic to the latter island break out, flourish and subside across the disputed territory where Heaney and other "Northern" poets have their roots' (1997: 1). In another essay in the same book, Edna Longley emphasises the role of language in particular for how Heaney addresses this theme, and suggests that *Wintering Out*, 'An aesthetic brand of revolutionary action, perhaps more linguistic reclamation than decolonisation, takes on the English language itself, with mixed declarations of love and war' (1997: 36). She refers for instance to the poem 'A New Song' (Heaney, 1972: 33):

> But now our river tongues must rise
> From licking deep in native haunts
> To flood, with vowelling embrace,
> Demesnes staked out in consonants.

In these lines, Irish and English are contrasted as soft, round vowels versus hard and sharp consonants, a contrast that recurs in several other poems.

Heaney's 'political' role as a poet arguably culminated with *North*, and the labyrinthine way in which he addressed that role has been discussed in Chapter 2. Harold Bloom suggests that Heaney's work before *North* had been both 'pastoral and personal' (2003: 21). The landscape characterised by violence that is depicted in *North* radically changed this, and signified a completely different, more complex relationship between his own personal past and that of Ireland's. Bloom states that, starting with *North*, 'The problem for Heaney as a poet henceforward is how not to drown in this blood-dimmed tide' (2003: 11).

Neil Corcoran notes further that one of the central tropes of *Field Work*, the collection immediately succeeding *North*, is 'a reciprocal relationship with Nature' (1998: 125). If the relationship between the bogs and the political landscape of Northern Ireland played out in the shadow of more striking images of contemporary violence and conflicts, in *Field Work*, Corcoran suggests that Heaney reflected more critically on the relationship between nature and culture. This chapter suggests that this reflection begins already in Heaney's earliest volumes, and continues through his entire poetics. Heaney has described this process himself, stating that the 'language and place-names in *Wintering Out*', especially poems like 'Broagh' and 'Anahorish', 'politicize the terrain' (Randall, 1979).

In the essay 'Against Authenticity: Global Knowledges and Postcolonial Ecocriticism', Cara Cilano and Elizabeth DeLoughrey point to a tendency to oversimplify the relationship between ecocritical and postcolonial themes:

> we want to suggest that too heavy a reliance upon constructing parallels between postcolonialism and ecocriticism can lead to an unproblematized division between people (on the postcolonial side) and nature (on the ecocritical one). To some extent, this parallelism between 'excluded, exploited, and oppressed' nature renders the two equivalent, thus dehistoricizing through natural and universal metaphors.
>
> (2007: 75)

This chapter attempts to show how Heaney resists this kind of simplification through the intricacies and interdependencies he creates in his poetics between colonial and ecological signifiers and ideas.

Wintering Out

The poem 'The Backward Look' (Heaney, 1972: 29–30) describes the flight and sounds of a snipe, intertwined with references to the Irish language. The poem balances the literal with the symbolical:

> A stagger in air
> as if a language
> failed, a sleight
> of wing.

These lines establish a theme that intertwines bird with language: the phrase 'stagger in air' might be read as a description of a snipe's interrupted flight, or as a reference to the sound of spoken words. In both cases, the word 'stagger' suggests that something is amiss. Taken as a reference to language, supported by the second and third lines, staggering refers to a 'language / failed', suggesting a language that is spoken without fluidity or confidence, disappearing into local dialect. Alternatively, read as a description of the flight of a bird, 'stagger' is complemented by the description 'sleight / of wing' at the end of the stanza. In this case, 'stagger' indicates an interrupted flight, perhaps of an injured bird. Read in this way, the single snipe in Heaney's poem represents not only itself but the snipe as a threatened species in modern Ireland (Parkin and Knox, 2010: 161).

The phrase 'sleight / of wing' also complicates the nature–culture relationship in this poem, as the word 'sleight' is usually used in a human context, to describe a 'sleight of hand'. Though animals can also have dialect, both language and dialect are primarily used when talking about humans. The words sleight, language and dialect as used in this poem with reference to the bird combine to describe the bird in human terms, and give it a kind of 'personhood'. In the sense that birds have dialects, the description of this particular bird's dialect acknowledges the relationship between species and place, and of an individual's relationship with place as something not limited to humans.

In the following two stanzas, as the snipe makes its own sound, the two possible readings of the first stanza merge:

> A snipe's bleat is fleeing
> its nesting ground
> into dialect,
> into variants,
>
> transliterations whirr
> on the nature reserves –
> *little goat of the air,*
> *of the evening,*
> *little goat of the frost.*

That the 'snipe's bleat' is 'fleeing its nesting ground' could be a good thing, suggesting simply that the speaker can hear the snipe. However, in this context it is perhaps more likely a reference to the fact that the bird is disappearing from its natural habitat.

In the second half of the second stanza, the zoomorphic bleat disappears 'into dialect, / into variants'. This shift from descriptions of the bird to a focus on language is important, as it makes it impossible to read the poem merely literally; without the reference to language the mentioning of dialect does not make sense. Comparing the sound of the bird to local dialects also furthers the suggestion that the bird is being forced to leave its original nesting ground and move into other, less suitable realms. These lines complicate the relationship between bird and language, as here the two themes inform each other in a direct, non-allegorical fashion. The three italicised lines refer to Irish names for the snipe, such as 'little goat of the frost' (the snipe, surprisingly, can sound like a goat, as is explained in lines 14–15 of the poem). If the snipe disappears completely from Ireland, these names, and thus parts of the Irish language, will likely disappear too.

The next stanza explains that the bleating sound actually comes from the bird beating its tail: 'It is his tail-feathers / drumming elegies'. The theme of extinction is enforced by the reference to the bleat as 'elegies', the suggestion that the snipe is following 'in the slipstream / of wild goose / and yellow bittern' (birds that are not common in Ireland today), and the description of the bird's gradual disappearance from sight, 'disappearing among / gleanings and leavings'. The reference to 'wild goose' further intertwines ecology with colonialism: 'The Wild Geese' is also a reference to soldiers who left Ireland to fight for other European countries between the sixteenth and eighteenth centuries.

Greg Garrard has drawn attention to the intertwined relationship between ecology and linguistics in 'The Backward Look', which makes it impossible to separate the declining snipe population and the marginalisation of Irish from each other. Garrard suggests that allegorical readings have distracted critics from the poem's literal referents; while language and colonialism have been discussed by several critics in relation to this poem, the snipe itself, according to Garrard, is 'not mentioned despite the brilliant infolded isomorphism of the fates of the snipe's own "language", the names it has in the Irish language, and the Irish language itself' (1999: 193).

The next stanza describes the snipe's 'flight / through the sniper's eyrie'. Garrard explains the complex references of the word 'sniper', and the prioritisation of the political over the ecological:

> The bird's 'flight / through the sniper's eyrie' links it with the contemporary Troubles, whilst also emphasising the very threatened liminality that make it an appropriate symbol of Gaelic: a 'sniper' is originally one who shoots snipe! Vehicle and tenor, nature and language, are so imbricated with each other here that the automatic and exclusive prioritisation of the latter would seem an obvious example of the perverse eco-blindness works such as *Romantic Ecology* and Lawrence Buell's *The Environmental Imagination* set out to challenge.
>
> (1999: 193)

The word sniper was originally coined by British soldiers in India in the early nineteenth century, and referred to a gunman skilled enough to shoot a snipe. In Heaney's poem, the word sniper is paired with the word 'eyrie', which refers to the nest of a bird of prey. The word 'eyrie' also suggests *Éire*, the Irish name for Ireland, which makes particular sense in the phrase of the snipe's 'flight / through the sniper's eyrie', transcribed as the snipe's flight through a colonised (Northern) Ireland, with the snipe representing at once itself, its species and the Irish language, while the sniper here suggests Britain. The colonial theme, associated with conflict and violence, can also be sensed in the words 'earthworks' and 'wall-steads', suggesting military fortifications or hiding places. By the end of the poem, the snipe gradually disappears from view, leaving behind an ambiguous collection of 'gleanings and leavings' from the previous line.

The tendency among literary critics towards what Garrard refers to as 'eco-blindness' is seen in several analyses of 'The Backward Look'. Michael Molino has suggested that the poem 'describes the deterioration of the Irish language as the flight of a snipe fleeing a hunter', notably using 'as' rather than 'and', which could have indicated a different, double meaning. Not noting the ecological references to the disappearance of the snipe as a reference to a declining snipe population, Molino suggests that the bird's gradual disappearance seems 'to invoke a pastoral vision, a coveted connection with or continuation of the true Irish tongue through those rural farmers who still speak Irish as their primary language'. Considering the image of the disappearing snipe as representative of a species, the description of the scene as pastoral is misleading. Molino suggests that the fieldworker in the last lines of the poem might be 'an archaeologist, a linguist, or poet whose archive is the "English" language, with all its dialects and variants, through which the snipe's flight might still be traced' (1993: 19). In addition, the 'fieldworker' might also be a biologist or ecologist collecting evidence of the actual snipe.

The poem 'Midnight' (Heaney, 1972: 35), also from *Wintering Out*, is discussed by Garrard in similar terms. 'Midnight' remembers the Irish wolf, extinct for about 200 years:

> Since the professional wars –
> Corpse and carrion
> Paling in rain –
> The wolf has died out
>
> In Ireland. The packs
> Scoured parkland and moor
> Till a Quaker buck and his dogs
> Killed the last one
>
> In some scraggy waste of Kildare.

As in 'The Backward Look', there is a relationship here between the extinction of a species and colonialism. However, in this poem the relationship from 'The Backward Look' seems to be at least partly inverted. Garrard notes that in the case of the wolf 'Not war but peace – in particular the habitat destruction that followed repressions of rebellions in 1601 and 1798 – killed off the Irish wolf'. Garrard quotes Patrick Sleeman, who points out that 'Since the impenetrable woods provided refuge for Irish rebels, priests and outlaws, as well as for wolves, their destruction, under the pretext of civilizing the country, made economic and political sense from the government's and colonist's point of view'. Garrard also notes

> a fairly close correlation between Acts of Union and the extermination of wolves: Wales was incorporated in 1536, and the last wolf sighting was in 1576; Scotland formed the United Kingdom in 1707, and the last wolf was seen there in 1743; Ireland joined in 1801, and the 'Quaker buck and his dogs' finished off the Irish wolf at about the same time
>
> (1999: 193–4, 195).

The correlation between military conflict and species extinction is more ambiguous in 'Midnight' than in 'The Backward Look', but seems to maintain a negative effect of a colonial presence on the native fauna.

The linguistic theme is not introduced in 'Midnight' until the last stanza, where it finally appears with a suddenness comparable to the introduction of contemporary victims into the context of ancient bog bodies in *North:*

Nothing is panting, lolling,
Vapouring. The tongue's
Leashed in my throat.

Garrard notes that in these lines 'the powerful non-presence of the wolf is linked to the "leash" of the oppressor's language – the language of the poem itself, of course' (1999: 195). At first sight, the description that 'Nothing is panting, lolling' refers to the absence of wolves in the landscape, but in light of the last sentence, it equally refers to the poet's own ability to speak, in the language of the coloniser. Writing his poem in a language that is in some ways not his own, the speaker compares himself to a tamed wolf 'crossed / With inferior strains'. If 'The Backward Look' suggests that the extinction of the snipe will contribute to a set of Irish words and expressions being forgotten, 'Midnight' seems to suggest that the successful repression of rebellion results in the extinction of a species.

Garrard also discusses the poem 'Serenades' (Heaney, 1972: 62), featuring a corncrake, as part of the same category of colonial/ecological poems. The speaker in this poem states that his voice has been like that of the corncrake:

My serenades have been
The broken voice of a crow
In a draught or a dream,
The wheeze of bats

Or the ack-ack
Of the tramp corncrake

Discussing this poem, Jonathan Allison describes the corncrake's sound as 'hoarse', suggesting that it may be seen as simultaneously representing a 'guttural muse' and 'the Anglo-Saxon poetic tradition', and furthermore that 'as such it represents a blend of both English literary tradition and native Irish experience'. Allison also argues that 'the crake might also be considered a symbol of Ulster's Catholics, marginalized and intimidated by Unionist hegemony' (1996: 74, 77). Allison refers to several poems by Heaney that mention the corncrake, and states that:

> In 'Serenades', [Heaney] expressed an ecological consciousness of the bird as victim of modern agricultural techniques, marginalized in an anti-pastoral landscape of 'combines and chemicals', while also constructing the crake as an emblem of the poet's 'guttural muse', rooted in the Ulster landscape but also alluding to the 'alliterative tradition' in its repetitive call. The bird's crepuscular call is closest to conventional idyllic pastoral in 'Glanmore Sonnets', but in 'Casting and Gathering' the call is mediated through the

mechanism of the reel and has become an integral part of the poet's dialectical voice.

(1996: 79)

As both Allison and Garrard note, 'Serenades' intertwines linguistic with ecological frames of reference, combining them in the complex figure of the corncrake. Garrard suggests that these poems highlight a need to read Heaney's poetry in a way that resists 'the peremptory subordination of environmental issues to the linguistic and political ones that dominate critical discourse' (1999: 196). A similar point is made by Lawrence Buell in *The Environmental Imagination*, where Buell stresses the importance of avoiding 'opposite reductionisms', defined as 'reductionism at the level of formal representation such as compel us to believe either that the text replicates the object-world or that it creates an entirely distinct linguistic world; and reductionism at the ideational level, such as to require us to believe that the environment ought to be considered either the major subject of concern or merely a mystification of some other interest' (1995: 13). Poems like 'A Backward Look' and 'Serenades' indicate some ways in which colonial/political and ecological/environmental themes can be intertwined in ways that avoid either of these kinds of reductionisms.

Human Chain

As in *Wintering Out*, references to the Irish language inform Heaney's last volume of poetry, *Human Chain*, from 2010. 'An Old Refrain' (Heaney, 2010: 20–1) develops the connection between language, ecology and colonialism suggested in the earlier poems. The poem begins:

Robin-run-the-hedge
We called the vetch –
A fading straggle

The 'we' established in these lines also implicates a 'they'. This opposition between 'us and them' suggests a colonial theme already in the first stanza. 'We' are defined by the name of a certain plant, in this case a common weed known by the name Robin-run-the-hedge. In this description, the common identity of 'we' is defined by its relation to and shared naming of a common plant, suggesting a shared sense of place with knowing of the local environment. The next stanza describes the straggle as resembling a 'Lincoln green / English stitchwork', introducing a link between the weed and the British, and a dichotomy between Ireland and Britain. 'Lincoln green' was a specific kind of dye used in England in the Middle Ages (specifically in the town Lincoln), and is particularly associated with the clothes worn by Robin Hood. The link between the vetch[1] and the legend of Robin Hood is supported by the word 'robin' in the name of the plant in the poem's opening line. The reference to Robin Hood strengthens the association between the vetch and the British, but it also complicates the relationship between Britain and Ireland through the association with a character of good deeds and noble intentions. The plant is then described as 'Unravelling / *With a hey-nonny-no*', suggesting that something is coming undone, especially when paired with 'fading' from the end of the first stanza. These two adjectives perhaps reflect the progress of the peace process in Northern Ireland, and an increasing independence in relation to Britain.

'*With a hey-nonny-no*' is the first of five italicised words or phrases in the poem. This expression appears in at least two plays by Shakespeare: in *As You Like It* it forms part of a song about two young lovers: '*With a hey, and a ho, and a hey-nonny-no, / These pretty country folks would lie*' (Act 5, Scene 3), while in *Much Ado About Nothing* it appears in a song that encourages women to let go of deceitful men, and convert all their '*sounds of woe / Into Hey, nonny nonny*' (Act 2, Scene 3). In Heaney's poem this phrase, like the reference to Robin Hood, suggests an English cultural and literary tradition. Other readings of Ireland as female and Britain as male (seen especially in *North*, in poems like 'Act of Union' and in several of the bog poems) make it possible to read this line as referring to the relationship between the British and Irish in terms of a male/female relationship. This ambiguous line might also be the 'old refrain' in the poem's title.

The following line describes the vetch unravelling 'Along the Wood Road', a reference to a road in the area where Heaney grew up. It is also the name of the next poem in *Human Chain*, 'The Wood Road' (Heaney, 2010: 22–3), which is a much darker poem about political violence in Northern Ireland. In this poem the Wood Road

> Resurfaced, never widened,
> The verges grassy as when
> Bill Pickering lay with his gun
> Under the summer hedge
> Nightwatching, in uniform –
>
> Special militiaman.

The 'hedge' in this poem's fourth line refers back to 'An Old Refrain' and the 'Robin-run-the-hedge'. 'The Wood Road' is more explicit in its references to the violent history of Northern Ireland; the speaker walking along the Wood Road in the poem remembers its past:

> that August day I walked it
> To the hunger striker's wake,
> Across a silent yard,
> In past a watching crowd
> To where the guarded corpse
>
> And a guard of honour stared.

Yet in its last line this poem also subscribes to the idea of 'unravelling', suggested in 'An Old Refrain':

Film it in sepia,
Drip-paint it in blood,
The Wood Road as is and was,
Resurfaced, never widened,
The milk-churn deck and the sign

For the bus-stop overgrown.

'The Wood Road' alludes both to memory and to the loss of memories. The references to film and painting suggest art as a creative form of remembering, while the description of the 'milk-churn deck' and the overgrown bus-stop seems to suggest the passing of time. The overgrown bus-stop is also reminiscent of the unravelling, overgrowing weed in 'An Old Refrain'.

'The Wood Road' (St Vincent Millay, 2004 [1923]: 8) is also a short poem by Edna St Vincent Millay that describes nature as a source of strength in times of grief, stating that even 'If I were to walk this way / Hand in hand with Grief',

  though Grief should know me hers
  While the world goes round,
It could not if truth be said
  This was lost on me:
A rock-maple showing red,
  Burrs beneath a tree.

The twice-named 'burrs' in this poem reappear in 'An Old Refrain', where the burrs also play an important role. In light of Millay's poem, perhaps 'nature' in 'An Old Refrain' can be read as a place for recovery and reconciliation.

The plant that the poem describes is notorious for its ability to cling or stick to things through fine, hooked bristles that cover all parts of the plant, including the burrs just mentioned, making it sticky like velcro. This ability has earned the plant nicknames such as stickyweed, stickyleaf and catchweed. This characteristic may also make it a symbol for colonialism, as suggested in the end of the first part of 'An Old Refrain':

Sticky entangling

Berry and thread
Summering in
On the tousled verge.

'Berry and thread' suggest a plant/cloth reciprocity also alluded to in the earlier reference to the vetch as 'A fading straggle / Of Lincoln green / English stitchwork'.

The word 'verge' can also refer to an area of particular jurisdiction; 'the tousled verge' that the vetch is 'Summering in / On' could then be read as Northern Ireland.

'Summering in' is the opposite of 'wintering out', the title of Heaney's earlier collection. The phrase 'wintering out' comes, as Heaney explains in *Stepping Stones*,

> from memories of cattle in winter fields. Beasts standing under a hedge, plastered in wet, looking at you with big patient eyes. Just taking what came until something else came along. Times were bleak, the political climate was deteriorating. The year the book was published was the year of Bloody Sunday and Bloody Friday.
>
> (O'Driscoll, 2008: 121)

The description of 'Summering in / On the tousled verge' can then be read as a reference to holding out in the midst of the Troubles, rather than moving away. In a different reading, the phrase 'summering in' may describe the approach of the British, and conveys a sense of threat, supported by the Latin name for this particular plant, *Galium aparine*, where the species name comes from the Greek word *aparo*, meaning 'to seize'. A third interpretation is possible if 'summering in' is read as the opposite to 'wintering out' in the sense that it refers not to waiting for better times but to those better times having arrived. This reading reflects that 'An Old Refrain' was written in times of an improving political situation in Northern Ireland, compared to the contemporary setting of *Wintering Out*.

This improvement is reflected elsewhere in Heaney's poetics, for example to a comparison between 'The Tollund Man', also from *Wintering Out*, and the later poem titled simply 'Tollund' (Heaney, 1996: 69), one of the last poems in the collection *The Spirit Level* from 1996. In the later poem, the speaker is once again in Jutland, where he has been before to see the actual Tollund Man. However, this time, instead of the rich, dark symbolism recorded in 'The Tollund Man', the poet finds, as he stands in the 'Jutland fields', something that 'could have been a still out of the bright / "Townland of Peace"', a place where 'Things had moved on'. Compared to the region described in the bog poems, where sacrifices are made for the sake of the community and in the name of the land, this time Jutland is a

> user-friendly outback
> Where we stood footloose, at home beyond the tribe,
>
> to make a new beginning
> And make a go of it, alive and sinning,
> Ourselves again, free-willed again, not bad.

The free will mentioned here perhaps suggests liberation from the archetypal pattern of violence described in the bog poems, enabling 'a new beginning'.

The brightness and sense of hope in 'Tollund' was occasioned by several specific events in the real world, as Heaney's elaboration on the background of this poem explains:

> The coincidence was extraordinary. The IRA announced the ceasefire on a Wednesday, the last Wednesday in August, if I'm not mistaken, and I was asked to write about it for the next weekend's *Sunday Tribune*. That same weekend I was also bound for Denmark, to do a reading in Copenhagen University, and inevitably I was remembering the visit I'd made to Jutland twenty-one years earlier, to see the Tollund Man. What happened, at any rate, was an unexpected trip to the actual bog in Tollund where the body had been found in the 1950s. [...] It was like a world restored, the world of the second chance.
>
> (O'Driscoll, 2008: 350–1)

In light of these lines, the reference in 'An Old Refrain' to 'Summering in' seems to carry a specific sense of hope, within the framework of Heaney's previous poetics.

In 2005, Heaney continues the story of both the Tollund Man and the improving situation in Northern Ireland in the poem 'The Tollund Man in Springtime' (Heaney, 2005). In this poem the speaker imagines that he is Tollund Man himself, stating that he is 'neither god nor ghost', he is 'simply lost / To you and yours' – he is aware that he was sacrificed by his own people 'for their own good', to secure the fertility of the land. Reawakening in present times and seeing how the natural environment has changed, he describes his own sacrifice as a response by his people

> to a sixth-sensed threat:
> Panicked snipe offshooting into twilight,
> Then going awry, larks quietened in the sun,
> Clear alteration in the bog-pooled rain.

The snipe from *Wintering Out* reappears here, in the Tollund Man's recognition of both a declining presence of birds in the landscape and perhaps the increasing acidity of the rain. However, the sense of hope and political improvement reflected by the progression from first 'The Tollund Man' to the bog poems in *North*, then to 'Tollund' and finally to this poem seems to signify a sense of hope with regard to the changes in the environment that the Tollund Man now registers, for even though

> I smelled the air, exhaust fumes, silage reek,
> Heard from my heather bed the thickened traffic
> Swarm at a roundabout five fields away
> And transatlantic flights stacked in the blue

He also recognises that

> Late as it was,
> The early bird still sang, the meadow hay

Still buttercupped and daisied, sky was new.

With determination and resolution, he rises from the ground, spits into his hands to rub off the dust of the ground 'in pollen's name and my own'. The next lines bring to mind Heaney's father and grandfather cutting turf in Heaney's perhaps most famous poem, 'Digging', from his first collection:

  As a man would, cutting turf,
I straightened, spat on my hands, felt benefit
And spirited myself into the street.

Counteracting the grim notion of history repeating itself stated so forcefully in the earlier bog poems, here the Tollund Man is able to assert that '"The soul exceeds its circumstances". Yes. / History not to be granted the last word'. This statement points to the successful political negotiations in Northern Ireland, emphasising the possibility of resolution of enduring conflicts.

Returning to *Human Chain*, the second part of 'An Old Refrain' moves focus, away from the vetch and towards the 'thingness' of other things, and especially the relationship between this 'thingness' and the name of the thing. This part of the poem is structured around four italicised, partly phonolexic words.[2] The first stanza states:

In *seggins*
Hear the wind
Among the sedge,

The word 'seggins' recalls the burrs in the poem's first part, as 'seggins' is another name for the common sedge *Sparganium ramosum*, also called burr reed. The burrs might be another allusion to Shakespeare – in *A Midsummer Night's Dream* Lysander exclaims: 'Hang off, thou cat, thou burr! vile thing, let loose, / Or I will shake thee from me like a serpent!' (Act 3, Scene 2). By choosing the vernacular 'seggins' instead of the more common name burr reed, these three lines emphasise ways of knowing the natural environment through local names for plants and places. The poem also builds on the contrast between Irish sounds as softer than English ones established in other poems, comparing the sound of 'seggins', where you can almost here the wind, to the different sound of 'burr reed'. Burr is also an Irish accent, and may refer to how language sticks to things, enforcing the connection between the disappearance of things and the disappearance of words established already in 'The Backward Look' and developed throughout Heaney's poetics.

In the next stanzas, the reader is encouraged to similarly 'hear' nature through the particular sounds of three more words:

In *boortee*
The elderberry's

Dank indulgence,

In *benweed*
Ragwort's
Singular unbending,

In *easing*
Drips of night rain
From the eaves.

'Boortree' is Ulster dialect for the elder tree. Discussing the poem 'Broagh' (Heaney, 1972: 27), in which the same word appears, Molino notes that 'boortree' is 'a word with a complex history in English and other languages' (1993: 192), probably with its origins in the Scottish pronunciation of 'bower tree'. The word also occurs in 'Broagh', and in the fifth part of the 'Glanmore Sonnets' (Heaney, 1979: 37), where it is already connected to the adjective 'dank', just as it is in 'An Old Refrain'. The sonnet establishes a link between the elderberry and violence, first in the description of the boortree's 'green young shoots, its rods like freckled solder' through the orthographical similarity between 'solder' and 'soldier', and second by the reference to 'Its berries a swart caviar of shot', described by David Lloyd as a 'double metaphor' that juxtaposes 'an image of delicate and rare caviar with the image of solid black shotgun pellets' (1981: 91). This ambivalent notion of violence is also present in 'Broagh', as noted by David-Antoine Williams, who suggests that 'the name "boortree" is connected with a kind of toy gun which is made from the wood of the tree' (2010: 113).

Lloyd suggests that, in the section of the 'Glanmore Sonnets' where the boortree appears, 'Heaney's use of densely textured language is a method of embodying the dense texture of the natural world to make the reader come into more complete contact with it', and furthermore that 'This progression reflects an assertion which is important in the corpus of Heaney's work: that as we heighten the quality of our interaction with and awareness of the natural world, we also heighten the quality of our human relationships' (1981: 92). This relationship is reasserted in 'An Old Refrain' through references that intertwine close attention to changes in the natural environment and how it is reflected in language and dialect with a notion of an improving political relationship between Ireland and Britain.

In the next stanza, '*benweed*' is a translation of the Irish name for ragwort, buachalán buí, also referred to as ragweed or bunweed. The ragwort is a poisonous wild flower (or weed) common in both Ireland and Britain. Like 'seggins', the word 'benweed' in this poem is an Irish word for a plant more commonly referred to by its English name. Molino notes that the way in which Heaney draws attention to Irish expressions in this and other poems is important for many reasons, not least because 'many Irish dialect terms exist in various local forms that are unknown to most people, even some Irish citizens, because there is no dictionary of Irish English'. Therefore, inserting Irish words 'into a poem written in English at once celebrates the oral tradition of Irish literature, broadens the scope of Anglo-Irish literature, and underscores the dynamics of Ireland's history and literature' (Molino, 1993: 192–3).[3] In this sense, the second part of 'An Old Refrain' is a tribute to Irish dialect and its particular relationship with the natural environment.

The word 'easing' in the final stanza is a shorter version of 'eavesing', in turn another word for 'eaves', the poem's last word. Eaves are the projecting corners of a roof that carry off rainwater (less commonly, it may also refer to the edge of a forest). Derived from 'eaves', the word 'eavesdrip', suggests drops of rainwater dispelled from the eaves. The eavesdrip is also the area around a building where water from the eaves falls. There is an ancient English 'law of Eavesdrip' according to which it was forbidden to construct a building in such a way that the water discharged from the eaves would affect the neighbouring property.[4] The phrase is also reminiscent of 'eave drops' from Coleridge's poem 'Frost at Midnight' from 1798. Together with 'unravelling' and 'unbending', 'easing' ends the poem on a positive note of improvement and letting up.

The italicised words of the second part draw attention to the characteristics of both the words themselves and their referents, pointing to a Heaneyesque fascination with the 'thingness' or 'quiddity' of things. In a review of *District and Circle*, 'The Dominion of the Physical: Reading Heaney', John Wilson Foster, also editor of *Nature in Ireland: A Scientific and Cultural History* (1997), suggests that Heaney's 'reader senses only a comforting assurance of the quiddity of the world, his poetry functioning as a kind of fodder, for the soul as well as the senses' (Foster, 2007). This sense of quiddity is an important theme throughout Heaney's ecopoetry, and is also a way of resisting allegory by insisting on the things themselves. Neil Corcoran notes something similar when he states with reference to part I of *Station Island* that 'the inanimate world insists a caustic, disciplinary presence; things become examples' (1987: 125).

In 'An Old Refrain', the notion of quiddity is combined with words with specifically Irish connotations, strengthening the relationship between the things themselves and their names in Irish dialects. In this sense, these words express a sense of ownership of both the English language and the natural environment. In a reading inspired by the connection to Millay's poem, there is a sense in which 'An Old Refrain' formulates a connection to the natural world in order to address a troubled cultural history. Lloyd suggests that:

> Certainly the single most apparent and destructive political and social reality of Northern Ireland is ancient and ongoing separation (social, cultural, religious and political). Heaney's most outstanding and important trait as a political poet is his ability to heal separation by placing us all back into contact with the meaningful parts of our lives and surroundings. Heaney attacks our acceptance of destructive bigotries and desires that separates us from our humanity and our world through his use of language and imagery which force the reader into intense and honest contact with the objects, sensations and figures.
>
> (1981: 90)

Lloyd's description suggests a way to read 'An Old Refrain', especially the second part that encourages the reader to attend closely to the relationship between particular words, their origins, and their natural referents. Different from 'The Backward Look' and 'Midnight', which also intertwine colonial and ecological references, 'An Old Refrain' sets up the relationship between nature and language in a way that encourages the reader to refer to it as a way to respond to, and perhaps even reconcile, the memories of the colonial history described in the first part.

The two parts of 'An Old Refrain' combine a kind of natural linguistics with what Molino refers to as Heaney's 'historical linguistics'. Starting with *Wintering Out*, Molino argues, Heaney began to develop

> a polyphonic voice that displaced the political and cultural antagonisms endemic to his country and relocated them in a realm of reflexive, historical linguistics. That is, the emancipating discourse that Heaney developed to circumvent Ireland's centuries-old and exclusive political monologues entailed a confrontation of historically diverse discourses within a single poem. Heaney chose to address the political/poetic dilemma that he and other Irish writers faced by making the language of Irish writers (as well as the language of Irish people in general) the focus of – or at least a significant factor in – his poetry. Heaney thereby circumvented the political/poetic dilemma with a poetry whose vernacular problematic addressed old antagonisms in an innovative way.
>
> (Molino, 1993: 181)

This 'historical linguistics' in Heaney's poetics is fundamentally intertwined with a natural linguistics, as poems like 'A Backward Look' and 'An Old Refrain' demonstrate, creating a simultaneous sense of both anchorage and slippage between reference and referent. Molino also notes that Heaney's poetry eventually comes to employ an array of words and expressions from multiple languages that combine 'a traditional English line of verse and create a form of poetry that circumvents political monologism by celebrating linguistic pluralism', through poems that 'enact the vernacular problematic of the Irish speaker/writer and explode in dialect and wordplay' (1993: 184, 188). This etymological wordplay and pluralism is similarly intertwined with referents to the history of the landscape, and to contemporary environmental change. Molino concludes his essay by stating that in *Wintering Out* Heaney finds a method to reinscribe 'Ireland's politics, literature, and languages' (1993: 201). Ireland's nature could be added to that list.

Notes

1 The poet/speaker calls Robin-run-the-hedge by the name of vetch but it seems that the plant in the poem is cleavers (goosegrass).

2 The term 'phonolexis' is suggested by Phil Roberts in *How Poetry Works*: 'There is no generally agreed name for this effect – meaning conveyed through phonemic

connotation limited to the speakers of a particular language. So I shall call it *phonolexis* (from the Greek words from "sound" and "vocabulary")' (2000: 54).

3 By 1998, five years after Molino wrote this, an Irish–English dictionary had actually been created.

4 *Oxford English Dictionary*. www.oed.com/view/Entry/59158?redirectedFrom=eavesdrip#eid [accessed Mar. 2014].

References

Allen, M. (ed.) (1997) *Seamus Heaney*, London: Macmillan Press.

Allison, J. (1996) 'Seamus Heaney's Anti-Transcendental Corncrake', in C. Malloy and P. Carey (eds), *Seamus Heaney: The Shaping Spirit*, pp. 71–81, Newark, DE: University of Delaware Press; London: Associated University Presses.

Bloom, H. (ed.) (2003) *Seamus Heaney*, New York: Chelsea House Publishers.

Buell, L. (1995) *The Environmental Imagination: Thoreau, Nature Writing, and the Formation of American Culture*, Cambridge, MA: Harvard University Press.

Cilano, C., and DeLoughrey, E. (2007) 'Against Authenticity: Global Knowledges and Postcolonial Ecocriticism', *ISLE*, 14/1: 71–87.

Corcoran, N. (1987) 'Seamus Heaney and the Art of the Exemplary', *Yearbook of English Studies*, 17, special number, British Poetry since 1945: 117–27.

Corcoran, N. (1998) *The Poetry of Seamus Heaney: A Critical Study*, London: Faber & Faber.

Foster, J.W. (2007) 'The Dominion of the Physical: Reading Heaney', *Pacific Rim Review of Books*, 3/1. www.prrb.ca/articles/issue05-heaney.htm [accessed Mar. 2014].

Garrard, G. (1999) 'Ecological Literary Criticism: Two Case Studies', doctoral dissertation, University of Liverpool.

Heaney, S. (1972) *Wintering Out*, London: Faber & Faber.

Heaney, S. (1979) *Field Work*, London: Faber & Faber.

Heaney, S. (1996) *Seeing Things*, London: Faber & Faber.

Heaney, S. (2005) 'The Tollund Man in Springtime', *Metre* (Spring).

Heaney, S. (2010) *Human Chain*, London: Faber & Faber.

Heise, U. (2010) 'Postcolonial Ecocriticism and the Question of Literature', in B. Roos and A. Hunt (eds), *Postcolonial Green: Environmental Politics and World Narratives*, Charlottesville, pp. 251–258, VA: University of Virginia Press.

Huggan, G. (2009) 'Postcolonial Ecocriticism and the Limits of Green Romanticism', *Journal of Postcolonial Writing*, 45/1: 3–14.

Huggan, G., and Tiffin, H. (2010) *Postcolonial Ecocriticism: Literature, Animals, Environment*, London: Routledge.

Lloyd, D. (1981) 'Seamus Heaney's "Field Work"', *Ariel*, 12/2: 87–92.

Malone, C. (2000) 'Writing Home: Spatial Allegories in the Poetry of Seamus Heaney and Paul Muldoon', *ELH*, 67/4: 1083–1109.

Molino, M. (1993) 'Flying by the Nets of Language and Nationality: Seamus Heaney, the "English" Language, and Ulster's Troubles', *Modern Philology*, 91/2: 180–201.

O'Driscoll, D. (2008) *Stepping Stones: Interviews with Seamus Heaney*, London: Faber & Faber.

Parkin, D., and Knox, A. (eds) (2010) *The Status of Birds in Britain and Ireland*, London: A&C Black.

Randall, J. (1979) 'An Interview with Seamus Heaney', *Ploughshares*, 5/3. www.pshares.org/read/article-detail.cfm?intArticleID=9476 [accessed Mar. 2014].

Roos, B., and Hunt, A. (eds) (2010) *Postcolonial Green: Environmental Politics and World Narratives*, Charlottesville, VA: University of Virginia Press.

St Vincent Millay, E. (2004) *The Harp Weaver and Other Poems*, Whitefish, MT: Kessinger [first published 1923].

Williams, D. (2010) *Defending Poetry: Art and Ethics in Joseph Brodsky, Seamus Heaney and Geoffrey Hill*, Oxford: Oxford University Press.

5 Ecosemiotics

Anti-anthropocentrism in Hughes's animal poems

Hughes's poem 'Brambles' (Hughes, 2003: 710–11) begins:

> The whole air, the whole day
> Swirls with the calls of jackdaws. The baby jackdaw
> Generation is being initiated
> Into jackdawdom – that complicated
> Court-world of etiquette
>
> And precedence, jingoism and law.

The idea of a 'jackdawdom' reflects the semiotic concept of *Umwelt*, defined by Jakob von Uexküll in 1920 as 'the self-centred world of an organism', or the world as it is subjectively perceived by different species or individuals (quoted from Cobley, 2010: 348). The concept of the *Umwelt* recognises that several species may share an environment and yet have very different experiences of that environment. The *Umwelt* of the jackdaws, as 'Brambles' suggests, is the world as it is known, understood and communicated by jackdaws, and each jackdaw must learn how to live and communicate according to the rules of their *Umwelt*. This means learning to interpret and respond to signs transmitted by other jackdaws, as well as by other species that influence and have meaning in the jackdaws' *Umwelt*. Describing the *Umwelt* of the jackdaws, 'Brambles' draws attention to the world as seen by the jackdaws, and to the existence of species-specific, other-than-human *Umwelten*, and to the limited perspective of the human *Umwelt*.

Ecosemiotics studies communication between humans and non-human nature by bridging biosemiotics (the study of non-human sign systems) and traditional or cultural semiotics (the study of communication between humans). This chapter reads 'Brambles' and other animal poems by Hughes that revolve around observation and interpretation of animals and nature from an ecosemiotic perspective, using theories and methods developed by ecosemioticians Kalevi Kull and Timo Maran. Two different kinds of ecosemiotic notions or practices in Hughes's poetics are addressed: The first is identified in poems that attempt to describe nature from a non-human perspective, that try to perceive other species' *Umwelten* and describe non-human nature in semiotic terms. 'Brambles' is an example of this kind of poem, and there are several others, especially in the two collections *Season Songs* (1976) and *Flowers and Insects* (1986). The second kind of poem is less focused on description and more concerned with interpretation, addressing how non-human nature is contextualised and acquires meaning in human *Umwelten*. This kind of poem is for example important in Hughes's collection *River* (1983), exemplified in this chapter by the poem 'Salmon Eggs', which points to some specific ways in which natural signs become meaningful for humans.

Ecosemiotics

Ecosemiotics is based on Peircean semiotics, which includes three main components: sign, object, and interpretant. The sign represents an object so that the object is brought into relation with the interpretant. A biosemiotic example is the relationship between flower, nectar and bee – the flower (the sign) represents the nectar (the object) to the bee (the interpretant) in such a way that the object and the interpretant are brought into connection with each other (the bee finds the nectar). This triadic view of the semiotic process is different from the semiotics of Ferdinand de Saussure, for example, who conceives of the same process dyadically, including only signifier and signified. Peircean semiotics, compared to a Saussurean version, emphasises the role of the interpretant. In Peircean semiotics, there are three different types of sign: iconic, indexical and symbolical. These three types relate to their objects in different ways – by similarity, physicality or habit, respectively.

From the basic definition as the study of sign systems, semiotics is divided into several subfields. Human sign systems are studied within the field of anthroposemiotics, more commonly referred to as cultural semiotics, or just semiotics. Non-human sign systems are studied in biological semiotics, or biosemiotics for short. Zoosemiotics is a subcategory of biosemiotics and refers to the study of animal communication, while phytosemiotics is the study of communication among plants. Ecosemiotics brings biosemiotics and cultural semiotics into contact with each other by studying communication between humans and their natural environment.

Ecosemiotics was first introduced in 1998 in two different articles in a special issue of the journal *Sign System Studies*, one by Kalevi Kull and another by Winfried Nöth. In the first one, Kull suggests that:

> *Ecosemiotics* can be defined as the semiotics of relationships between nature and culture. This includes research on the semiotic aspects of the place and role of nature for humans, i.e. what is and what has been the meaning of nature for humans, how and in what extent we communicate with nature. Ecosemiotics deals with the semiosis going on between a human and its ecosystem, or a human in one's ecosystem.
>
> (1998: 350)

Kull points out that, while both biosemiotics and ecosemiotics research semiotic activities in nature, unlike biosemiotics, ecosemiotics includes a historical aspect, studying how the relationship between different natures and cultures has developed over time:

> Ecosemiotics describes the appearance of nature as dependent on the various contexts or situations. It includes nature's structure as it appears, its classification (syntactics); it describes what it means for people, what there is in nature (semantics); and it finds out the personal or social relation to the components of nature, which can be one's participation in nature (pragmatics). In all this, it includes the role of memory and the relationships between different types of (short-term, long-term etc.) memory in culture.
>
> (1998: 351)

Kull categorises nature–human interaction into six different categories: (1) recognition and control, (2) decontextualisation, (3) operation and remodelling, (4) opposition and reduction, (5) understanding and devaluation, and (6) selfing and valuation. Each category requires interpretation, and moving between different categories involves both interpretation and translation, not least if more than one interpretant is involved, as the same sign or object can mean different things in the *Umwelten* of different individuals. Kull refers to Uexküll's notion that the *Umwelt* 'emphasised that every organism has its own subjective environment, which is different from any other, and in the case of different species of animals these differences can be very large' (Kull, 1998: 354). He also notes that such differences may also take place within the same species' *Umwelt*, based on different interpretations of different individuals, in relation to past experiences and different contexts.

For humans, for example, the notion of 'nature' has many different meanings. In an attempt to clarify these, Kull identifies four different categories: 'zero nature' refers to wilderness unknown by people; 'first nature' is nature as it is perceived by humans; 'second nature' is any nature that has been changed by humans (managed forests, modified species, parks, gardens and so on); and 'third nature' refers to images of nature informed by the image of nature as perceived by humans ('first nature'), including aesthetic representations of nature in poems and paintings as well as scientific models. Kull schematises how these different natures and images of nature inform each other in this way:

0 – zero nature is	– nature from nature
1 – first nature is	– image from nature
2 – second nature is	– nature from image
3 – third nature is	– image from image

So, in summary:

> Zero nature is that which biologists want to describe. The first one is that which they perceive and describe. The second one is the one in their lab. And the third nature is what they get in their papers and models.
>
> (Kull, 1998: 357)

Kull acknowledges that these categories might seem trivial, but maintains that failing to recognise them continually leads to mistakes as we fail to recognise complex relations between nature and culture.

In the second article in the same special journal issue, Winfried Nöth defines ecosemiotics as 'the study of the *semiotic* interrelations between organisms and their environment' (1998: 333). This definition is slightly different from Kull's, who states that ecosemiotics is concerned specifically with the overlaps of biological and cultural semiotics – Nöth's definition suggests a more relational study of organisms and their environments. While in Kull's defintion, eco- and biosemiotics are two different (though related) fields of study, Nöth seems to suggest that ecosemiotics is a subfield to biosemiotics. In a later essay, however, Nöth approaches Kull's definition and suggests that ecosemiotics is

> situated between the semiotics of culture on the one hand and the semiotics of nature on the other. *Culture* is involved since models developed in cultural history determine the way in which humans interpret their natural environment. Nature is involved not only since our own natural environment is the object of ecosemiotic research, but also since the orientation of organisms in prehuman life equally involves environmental semiosis.
>
> (2001: 71)

Differentiating between two-way communication and one-way signification, he adds that

> *Communication*, defined as a sign process that involves a sender and a receiver, occurs not only among humans, but also between all other organisms throughout the whole biosphere. Not only cultural semiotics, but also bio- and zoosemiotics are hence concerned with processes of communication. *Signification*, by contrast, which concerns sign processes without a sender, predominates in ecosemiotics, where organisms interact with a natural environment that does not function as the intentional emitter of messages to the interpreting organism.
>
> (2001: 72)

The distinction between intentional and unintentional communication, or between communication and signification, points to how humans attribute meaning to signs that are not intended for them. Examples include birdsong, which for humans may indicate an idyllic pastoral setting, but may actually be the expression of territorial claims. Although the signs/songs are not intended for humans, they nevertheless acquire meaning in human *Umwelten*.

Nöth notes that ecosemiotics presumes a very low threshold for what qualifies as a sign. That presumption is, as Nöth points out, very different from twentieth-century structuralist semiotics, which considered non-lingual thought and communication 'a shapeless and indistinct mass' (Ferdinand de Saussure, quoted by Nöth, 2001: 72). Noting that structuralist semiotics has studied the natural world as a background structure in different texts, Nöth suggests that this type of study of nature as culture could be described as a specifically cultural subfield of ecosemiotics. He suggests four different frameworks through which nature is 'culturalised', including magical, mythological, metaphorical and pansemiotic. This categorisation can be used to define Hughes's mythological depictions of nature and environment, for example, as part of a tradition of instructive texts through which ecological knowledge has been transmitted between generations and cultures, informing people about their place in and relationship with their surroundings.

Nature-text

Building mainly on Kull's definition and development of ecosemiotics, Timo Maran has suggested the concept of the 'nature-text' to develop ecosemiotics into a practical methodological tool that can take into account and integrate biological and cultural semiotics, so that the resulting analyses address both the semioticity of nature and its depiction in written texts. Maran's methodology is thus an attempt to include the kind of cultural ecosemiotics that Nöth describes while maintaining a strong connection to biosemiotics, describing the relationship between 'the written text and the natural environment', and drawing attention to 'nature writing as an appreciation of an alien semiotic sphere' (Maran, 2007: 269). Providing a systematic framework for analysis of the meaning of different parts of the natural environment, this methodology also offers a way to compare interpretations of natural signs in different cultural and historical contexts, to highlight differences and similarities.

Understanding the transformation of 'first nature' (images of nature) into 'third nature' (images of images of nature) requires different models for mediation and translation to be connected. The concept of the nature-text presents a methodology for doing this by integrating perspectives from bio- and cultural semiotics to highlight representations of semiotic activities in the natural world, as well as semiotic relationships between nature and culture. The most obvious study object for his methodology, Maran suggests, is nature writing, defined as 'a written text that is related through meaning relations to a part of natural environment' (2007: 287). Understood as a form of nature writing, interpretation of so-called 'nature poems' then depends on analysis of both intertextual and contextual elements. Maran's statement that 'The understanding of nature writing does not depend solely on interpretation of the written text, but also on structures of outer nature, which have their own memory, dynamics and history, and if those outer structures change, then the field of possible interpretation for the written text will also change' (2007: 280) connects to Heaney's comment quoted in the introduction to this book, that 'environmental issues have to a large extent changed the mind of poetry'. Both Maran and Heaney point to how changes in the outer environment change how we interpret texts, including fiction, non-fiction and poetry, about that environment. The concept of the nature-text that Maran suggests is thus twofold, comprising the written text as well as the natural environment.

Maran's methodology is developed within the frames of the Tartu-Moscow semiotic school, which understands the concept of text as including anything with a cultural meaning and value. According to this tradition, the environment itself may be read as a text, which implies the presence of agency in the natural world. To explore this notion, Maran draws on the work by zoosemiotician Thomas Sebeok, who argued in 1988 that before language there is another, preceding system, which he calls 'the-world-as-perceived'. Interested in the semiotic instinctual relationship between organisms and their environment, Maran employs Sebeok's work to understand the mechanisms that allow different organisms to recognise and distinguish between signs, and translate them into changes and reactions in their behaviour. Humans also interact with their surroundings in this way, but in contrast to other organisms, we also have a verbal and conscious modelling system. Maran adopts Sebeok's differentiation between these two different systems, 'the anthroposemiotic verbal, which is unique to the human species, and the zoosemiotic nonverbal, which unites us with the world of nonhuman animals':

> The existence of a primary zoosemiotic modeling system is hard to notice for humans, because we are born into it (which makes it self-evident) and also because it is to a large extent overwritten by the system of conventional meanings. The existence and properties of the-world-as-perceived become, however, more apparent if the perceptual possibilities and communication systems of different species are studied. Direct and spatial perceptions, tactile and olfactory sensations as well as many occurrences of nonverbal communication between humans belong to the sphere of nonverbal modeling. Language resources are often insufficient for describing these kinds of phenomena, but it is certainly possible (and this is often done) to express these kinds of sensations by textual means.
>
> (Maran, 2007: 286)

In 'Brambles', Hughes draws attention to this zoosemiotic process in humans as well as in other organisms, emphasising similarities and kinship between human and non-human nature and promoting the awareness of humans as zoosemiotic organisms, countering our prevailing awareness of ourselves as 'anthroposemiotic' organisms.

Alien semiotic spheres

'Brambles' dramatises sign activity in non-human nature. The first stanza indicates the *Umwelt* of jackdaws, describing young jackdaws learning the rules of their world, in the lines quoted in the beginning of this chapter:

> The whole air, the whole day
> Swirls with the calls of jackdaws. The baby jackdaw
> Generation is being initiated
> Into jackdawdom – that complicated
> Court-world of etiquette
>
> And precedence, jingoism and law.

The 'world of etiquette' describes signs transmitted and interpreted by jackdaws; introducing the younger birds into 'jackdawdom' means teaching them to interpret and transmit those signs so that they can understand and be understood within the world as perceived and communicated by jackdaws. The next stanza emphasises the jackdaws' confinement and inability to move outside of this *Umwelt*:

> Nearly a prison world – with bars
> Of cries and signals. The jailors
> Are all the other jackdaws.

The next stanza moves from the animal to the plant world, describing a set of briars, asking:

> Briars are such a success, their defences
> So craftsmanlike,
> Their reachings so deliberate, are they awake?

The question seems to suggest an anthropocentric interpretation of the briar's intention. However, this notion is reversed in the next lines:

> Surely [the briars] aren't just numb,
> A blind groping. Yet why not?
> Aren't my blood-cells the same?
> What do even brain-cells fear or feel
> Of the scalpel, or the accident?
> They too crown a plant
> Of peculiar numbness.

This stanza, and the following lines that state 'brain-cells [...] too crown a plant / of peculiar numbness', depart from the distinction between zoosemiotics and anthroposemiotics made by Sebeok, and suggest that the difference between these two systems is perhaps not as definitive as we often perceive it to be. In these lines, the question moves from whether the briars are awake, to whether anything or anyone is awake. The next stanza emphasises the connection between the plant and the animal world:

> the jackdaws
> Work darkly to be jackdaws
> As if they were seeds in the earth.

These lines also relate to Kull's fourth category of processes involved in humans' perception of nature, 'opposition and reduction'. Kull suggests that recognition means making distinctions, such as between nature and culture. However, once the distinction has been made, it leads us to perceive the distinction as existing by itself, not as something we impose on the outside world. The distinction can also include reduction: defining nature and culture as opposing realms can make it more difficult to perceive the ways in which they interact, and ways in which they are similar. By questioning distinctions between humans, plants and animals, 'Brambles' encourages the reader to pay attention to and rethink the anthropocentric perspective inherent in those categories by first suggesting that plants are like humans (the briars are 'awake'), and then that humans are like plants (human brain cells are 'numb', just like the briars), and then that animals are also like plants (the jackdaws are like seeds).

The next stanza suggests that the world is infused with semiotic activity – even in the organism's *Innenwelt* (referring to an organism's inner environment, as compared to the outer environment referred to by the concept of *Umwelt*) 'mute cells' interpret and transmit signs, without awareness or recognition of the organism of which they form part:

> The whole claque is a benighted religion
> Around the godlike syntax and vocabulary
> Of a mute cell, that does not know who we are
> Or even that we are here,
> Unforthcoming as any bramble flower.

The syntax of the cell is neither more nor less 'godlike' than human language, suggesting at the same time that the communication in and between cells is as surrounded by a pansemiotic or mythical framework as human language. Further undermining or questioning the distinction between nature and culture by immediately reversing the suggestion that cells are like humans, the last of these lines suggests that the cells do not communicate at all, and are as quiet, as 'unforthcoming', as a flower.

The poem 'Tern' (Hughes, 2003: 720), also from *Flowers and Insects*, like 'Brambles' describes the world of a bird through linguistic terminology. It depicts the tern responding to signs in its environment in an immediate, unselfconscious way. The wings are 'remote-controlled / By the eyes', so that when the eyes see something, the tern reacts without hesitation:

> Suddenly a triggered magnet
> Connects him downward, through a thin shatter,
> To a sand-eel. He hoists out, with a twinkling,
>
> Through some other wave-window.

The bird's eye and its brain are described as 'a gimlet', waiting in the sky for a sign to appear below:

> He hangs,
>
> A blown tatter, a precarious word
> In the mouth of ocean pronouncements.
> His meaning has no margin.

Leonard Scigaj notes that 'Tern' illustrates 'the intimate symbiosis of bird or insect with the ecology of its environment' to the extent that the bird can be considered 'the ocean's utterance' (1994: 176), in the sense that its actions are immediate responses to triggers in its *Umwelt*; when the bird's eye detects a movement in the water beneath him, the reaction is immediate. The line stating that the bird's 'meaning has no margin' is key, and refers to the fact that there is no separation or distance between the sign and the bird's reaction. This is a recurring theme in Hughes's poetry, and is contrasted to human contexts where there is doubt, self-consciousness, consideration of alternatives, and so on. For the bird, there is room for no such things. For Hughes this human margin between sign and reaction is often something negative, suggesting that humans have lost an intuitive, instinctual relationship with their environment, so that they not only interpret but also misinterpret natural signs, and become alienated from their natural settings.

For Hughes, one way for humans to reconnect with their natural environment is through creative acts, including and even most importantly, the act of writing poetry. The description of the tern's flight, its attention to and symbiosis with its surroundings, and the sudden capture of its prey can all be read metaphorically as Hughes's descriptions of poetic creation. He explains his views on the relationship between capturing real animals and writing animal poems in the essay 'Poetry in the Making', in *Winter Pollen*:

> In a way, I suppose, I think of poems as a sort of animal. They have their own life, like animals, by which I mean that they seem quite separate from any person, even from their author, and nothing can be added to them or taken away without maiming and perhaps even killing them. And they have a certain wisdom. They know something special … something perhaps which

> we are very curious to learn. Maybe my concern has been to capture not animals particularly and not poems, but simply things which have a vivid life of their own, outside mine.
>
> (Hughes, 1994: 10)

In Hughes's view like the bird in 'Tern', which forms a perfect, unselfconscious ecological unit with its surroundings, the words and images of a successful poem relate immediately and powerfully to the imagination as well as to the physical environment. This notion intertwines a biosemiotic with a cultural semiotics perspective. Similarly to the description of the bird as an utterance of the ocean, the successful poem can be described as an utterance of the imagination; similarly to how the bird catches the sand-eel, the poem catches a thought or feeling from at the same time the inner and outer world:

> How can a poem, for instance, about a walk in the rain, be like an animal? Well, perhaps it cannot look much like a giraffe or an emu or an octopus, or anything you might find in a menagerie. It is better to call it an assembly of living parts moved by a single spirit. The living parts are the words, the images, the rhythms. The spirit is the life which inhabits them when they all work together. It is impossible to say which comes first, parts or spirit. But if any of the parts are dead, if any of the words, or images or rhythms do not jump to life as you read them, then the creature is going to be maimed and the spirit sickly.
>
> (Hughes, 1994: 12)

Compared to other human activities, where for Hughes there is a space between action and meaning, poetry has more in common with principles of ecology and myth (which for Hughes are closely related) than with modern, Western cultures and societies. In the perfect poem, there is no margin between the imagination and the poetic expression. The idea that poetry functions according to essentially ecological principles is a central thought in Hughes's ecopoetics.

If 'Brambles' and 'Tern' describe animals in perfect harmony with their environments, or *Umwelten*, 'Wolfwatching' (Hughes, 2003: 754), from the collection with the same name (1989), suggests what happens to a species when it is taken out of its natural habitat. The poem describes a wolf in a London zoo that, removed from its *Umwelt*, is reduced to 'listening to London'. As a result, the wolf is bored and sluggish, a mere shadow of a 'real' wolf:

> Woolly-bear white, the old wolf
> Is listening to London. His eyes, withered in
> Under the white wool, black peppers,
> While he makes nudging, sniffing offers
> At the horizon of noise, the blue-cold April
> Invitation of airs. The lump of meat
> Is his confinement. He has probably had all his life

Behind wires, fraying his eye-efforts
On the criss-cross embargo.

The discrepancy between the wolf's natural surroundings and the experiences of this individual wolf (such as the sounds of London) illustrates its entrapment between two different semiospheres: one that it is physically in but unable to interact with, and one that it has evolved to respond to. Compare, for example, the bee in the poem 'The Honey Bee', which is described as

Brilliant as Einstein's idea
Can't be taught a thing.
Like the sun, she's on course forever.

As if nothing else at all existed
Except her flowers.

The wolf, in comparison, is off course, out of tune with its environment, as suggested in the lines describing the mismatch between the wolf's sensory apparatus and the signs that surround it, which then become indecipherable. As a result, the wolf's surroundings have 'worn him away', and made it lose its 'wolfdom'. Rather than a real wolf, it has becomes more like 'an old man' or a toy, 'a woolly play-wolf'. These anthropomorphic descriptions suggest that the wolf has lost its function and role in its own species, *Umwelt* and instead become a sign in a human semiosphere.

This displacement of an object or sign is described by Kull's second category of nature–human interaction that describes decontextualisation, building on the previous category of recognition and distinction. Using the example of relocation of plants, Kull states that

> Recognition of an object, at least to some extent, decontextualises it. To be able to replant specimens of a useful species automatically means that specimens are taken out of their original biocoenosis, removing with this many connections with other species which are not taken with the plant to its new habitat. In built (artificial) ecosystems, for instance fields or parks, people often grow foreign, non-indigenous plants. Therefore, species in such places may encounter new relationships with other species which they have never experienced before. These plants are taken away from the context in which they have evolved.
>
> (Kull, 1998: 353)

'Wolfwatching' illustrates these processes. The zoo is a built artificial ecosystem, where the wolf is unable to relate to surroundings that do not correspond to his natural environment or senses.

The anthropomorphising of the wolf, the comparison of the animal to an old man or a toy can also be understood through Kull's third category of nature–human interaction, of 'operation and remodelling'. This category involves human remodelling of nature into a form that makes it recognisable and gives it a certain meaning in a human *Umwelt*. According to Kull,

> Operation always depends on (is regulated by) the forms and images the organism (a human) has acquired. Operation does not follow the whole structure of the environment and its webs of relationships, but discretisises it and disregards many sides. Consequently, it changes the environment, making it more similar to the human's own face.
>
> (1998: 353)

This process occurs in Hughes's poem, but it also takes place on the level of the zoo as a phenomenon, to which the poem also refers.

Decontextualised and anthropomorphised, the weight of the wolf's head is 'useless': it has no purpose and no relation to its surroundings:

> All his power is a tangle of old ends,
> A jumble of leftover scraps and bits of energy
> And bitten-off impulses and dismantled intuitions.

Unlike the baby jackdaws that are introduced into their *Umwelt* in 'Brambles', the displaced wolf 'no longer / Knows how to live up to' its own being, its 'wolf-pelt'.

The failure of the old wolf is contrasted by a younger individual, whose faculties are 'still intact', still 'waiting / For the chance to live':

> The rufous ears and neck are always ready.
> He flops his heavy running paws, resplays them
> On pebbles, and rests the huge engine
> Of his purring head. A wolf
> Dropped perfect on pebbles.

Contrasting the initial hopefulness and energy in the first of these lines, the next lines suggest the uselessness of his readiness, describing him as 'A product / Without a market'; the younger wolf is as decontextualised as the older one, its faculties have just not yet been worn away. It seems that in time, the younger wolf will suffer the same destiny as the old one, as the 'iron inheritance' of evolution becomes increasingly frustrated:

> The incredibly rich will, torn up
> In neurotic boredom and eaten,
> Now indigestible. All that restlessness
> And lifting of ears, and aiming, and re-aiming
> Of nose, is like a trembling
> Of nervous breakdown, afflicted by voices.

The restless movements of the wolf's ears, the 'aiming' of its nose indicate a search for signs that will direct its behaviour: 'Is he hearing a deer? Is he listening / To gossip of non-existent forest?' The description points to the persistent relationships between signs, objects and interpretant in the ecosystem as well as a discrepancy between different time scales – the relocation of a few single wolves does not affect their sensibilities as they have been shaped by evolution, leaving them debilitated and disabled in their new, constructed environment.

River

Kull's and Maran's theories of ecosemiotics address the meaning of nature for humans, and communication between human and non-human nature. While *Flowers and Insects* and *Season Songs* draw attention to the semiotic relationship between non-human species and their environments, Hughes's collection *River*, from 1983, can be read as a reflection on how non-human nature acquires meaning in human *Umwelten*.

Terry Gifford describes *River* as 'the last themed collection of [Hughes's] own work to engage directly with humans' relationship with the forces of the natural world' (2009: 56). The poems in *River* depict a spiritual global ecosystem, stretching from the sky to the sea – 'December River' (Hughes, 2003: 339–41) states that 'This vein from the sky is the sea-spirit's pathway'. The collection follows a river for one year, from winter through spring, summer and autumn and back to winter; from 'The Morning before Christmas' to 'Four March Watercolours' to 'An August Salmon' and 'October Salmon', before concluding with the 'January haze' of the last poem, 'Salmon Eggs' (Hughes, 2003: 639–81). Gifford notes that *River* illustrates Hughes's consideration of 'the symbolic value of a river as a "vein" in the life of the "sea-spirit" that regulates our globe'. For Hughes, the river is a sign that functions as 'a key indicator of the state of our relationship with our home' (Gifford, 2009: 57).

The most important presence in *River*, next to the actual river, is salmon. While the river signifies the interconnectedness of the global ecosystem, the salmon, framed by religious imagery, signifies the sacredness of that ecosystem. This is most explicit in 'Salmon Eggs', which frames the return of the salmon to its birthplace in terms of a religious ceremony. Looking into the water, the speaker sees and interprets the foreign world of the bottom of the river:

> I make out the sunk foundations
> Of dislocated crypts, a bedrock
> Time-hewn, time-riven altar. And this is the liturgy
> Of Earth's advent – harrowing, crowned – a travail
> Of raptures and rendings. Perpetual mass
> Of the waters
> Wells from the cleft.

Gifford notes that in these lines life and death are 'delicately held in juxtaposition', forming an 'elemental momentum' that functions as a sign of the regenerative process of the earth, where life and death form part of a healthy system – as also depicted in the *Elmet* poems (discussed in Chapter 3). In a sense, these poems combine the depiction of the regenerative processes of the natural world in the *Elmet* poems with the signification of that world as sacred and spiritual as portrayed in *Crow* (discussed in the first chapter of this book). Just as Crow is unable to comprehend the ocean in 'Crow and the Sea', the speaker in this poem is unable to comprehend or interpret the significance of the river; in the last line, his 'mind condenses on old haws'.

The speaker tries to submerge himself in the river and see it from a point of view that is not his own, attempting to perceive the river as something other than a reflection of his own *Umwelt*:

> In bone damp cold
> I lean and watch the water, listening to water
> Till my eyes forget me
> And the piled flow supplants me, the mud blooms
> All this ponderous light of everlasting
> Collapsing away under its own ephemera

As the last line suggests, he fails to transcend his own *Umwelt*, and instead he interprets the salmon in a cultural context by turning it into a sign that indicates a sacred cycle of life and death, also represented by the river:

> Sanctus Sanctus
> Swathes the blessed issue.
> Perpetual mass
> Of the waters

The religious terminology in the poem can be read in two ways; it might suggest the difficulty of moving outside a Christian framework for describing something, in this case the natural world, as sacred, a question also grappled with in *Crow*. Such a reading would suggest that it is this difficulty of moving outside one's cultural tradition that in the poem's last line causes the mind to condense 'on old haws'. Alternatively, the religious terminology can be read as an attempt to replace Christian concepts with ideas based on a spiritual view of the natural world, relocating the spiritual world from the otherworldly to the natural environment, a process that is also played out in *Crow*. Referring to the river as 'Sanctus Sanctus', this reading suggests that Hughes attempts to change the meaning of that phrase from signifying a Christian concept to instead signifying the flow of water, a natural instead of a transcendental phenomenon. According to this interpretation, a distinctively Christian sign, the 'Sanctus Sanctus', is inscribed in the poem with the intention to change its meaning and interpretation.

The salmon indicates the river as an iconic sign, meaning that it is based on the similarity between the circularity of its life journey and the similar circularity of the water cycle. It is also an indexical sign, as it is physically part of the river.

Moreover, it is a symbolical sign associated with the various meanings that have been assigned to salmon in different cultures. The mythology of salmon allows Hughes to use it as a shamanic device, as noted by Ann Skea, who suggests that, while the river itself is described as sacred,

> The Goddess is embodied, too, in the plants and many of the river creatures. But the river creatures, more than anything else, act as Shamanic guides for Hughes, drawing him through the sliding 'water mirror' meniscus of the river's surface into the deep, fluid Otherworld of imagination. Salmon, in particular, serve this function in *River*, and salmon were always regarded with special reverence by the Celts as being the source of wisdom and inspiration. (1994)

The nature-text of 'Salmon Eggs' recognises the life cycle of the salmon and suggests a cultural interpretation of the river as sacred based on the relationship between the circularity of the water cycle of which the river forms part and the life journey of the salmon, described as 'More vital than death', as 'death here seems a superficiality'. The poems take part in this sacred enactment of natural processes by reflecting the circular pattern in the structure of the collection, moving from January to January. While the river and salmon are iconic signs that represent cyclical processes that in turn indicate something 'more vital' than life and death, the life cycle of the salmon and the journey of both salmon and water are portrayed as signs that indicate an earth goddess, or nature as sacred. The poem is an interpretant of these signs, and the vacillations between the speaker trying to move outside his own *Umwelt* on the one hand, and the cultural framing of the river and salmon drawing on religious imagery on the other hand intertwine the idea of river and salmon as they exist outside the poem with their interpretation and role in human cultures and mythologies.

By signifying the river as sacred while at the same time acknowledging its otherness, in the last lines of the poem, 'Salmon Eggs' is also a nature-text that, in accordance with Maran's definition, can be 'used to describe nature writing as an appreciation of an alien semiotic sphere' (Maran, 2007: 269). While 'Brambles' and 'Tern' exemplify how Hughes's poems can be read as nature-texts that draw attention to semiotic acts in non-human nature and show how poems can be interpreted as nature-texts that endeavour to raise 'natural foreign semiotic spheres above the interpretation threshold of human culture' (Maran, 2007: 288), the *River* poems highlight how natural signs are transformed into cultural signs, sometimes in ways that separate them from their original context, but sometimes also in ways that create links between the different *Umwelten* of humans and other species.

References

Cobley, P. (ed.) (2010) *The Routledge Companion to Semiotics*, London and New York: Routledge.

Gifford, T. (2009) *Ted Hughes*, London and New York: Routledge.
Hughes, T. (1994) *Winter Pollen: Occasional Prose*, London: Faber & Faber.
Hughes, T. (2003) *Collected Poems*, London: Faber & Faber.
Kull, K. (1998) 'Semiotic Ecology: Different Natures in the Semiosphere', *Sign System Studies*, 26: 344–71.
Maran, T. (2007) 'Towards an Integrated Methodology of Ecosemiotics: The Concept of Nature-Text', *Sign Systems Studies*, 35/1–2: 269–94.
Nöth, W. (1998) 'Ecosemiotics', *Sign System Studies*, 26: 332–43.
Nöth, W. (2001) 'Ecosemiotics and the Semiotics of Nature', *Sign System Studies*, 29/1: 71–81.
Scigaj, L. (1994) 'Ted Hughes and Ecology: A Biocentric Vision', in K. Sagar (ed.), *The Challenge of Ted Hughes*, pp. 160–181, New York: St Martin's Press.
Skea, A. (1994) 'Ted Hughes and the British Bardic Tradition', Symposium Paper, University of Cairo, Dec. http://ann.skea.com/cairo.htm [accessed Mar. 2014].

6 'The place in me'
Heaney, globalisation and sense of place

In an anniversary article for 'Heaney at 70', Dennis O'Driscoll states that

> In his 70 years, Seamus Heaney has lived through many eras, growing up in rural Derry, where blackberry-picking was an annual rite, and entering the age of the BlackBerry smartphone. His first light flickered from candles and paraffin lamps; by 1995, the Nobel Foundation's spotlight had beamed in his direction. From the present age, when spring water is commercially bottled and marketed, he can look back on a childhood in which household water was carried from iron pumps and stone wells. Having been all ears for the crackling voices from a wet battery wireless in the 1940s, he has lived to record each of his poetry collections on CD.
>
> (O'Driscoll, 2009)

In 'the Heaney era', technological developments, globalisation and environmental changes have changed both the world we live in and how we view it. Global perspectives have come to complement and sometimes replace local considerations. This change between local and global does not only affect environmental conditions and concerns; Heaney has for example seen a developing peace process in Northern Ireland, which has also been a global process involving mediators and negotiators from international institutions and agencies.

In *Sense of Place and Sense of Planet: The Environmental Imagination of the Global*, Ursula Heise argues that globalisation is now

> the central term around which theories of current politics, society, and culture in the humanities and social sciences are organized. In literary and cultural studies, it is gradually replacing earlier key concepts in theories of the contemporary such as 'postmodernism' and 'postcolonialism'.
>
> (2008: 4)

In her study, Heise examines how a sense of belonging to local surroundings translates, or does not translate, into a sense of global responsibility and place attachment. The ability to translate the local into the global is perhaps especially relevant in relation to environmental issues, as reflected, as Heise notes, in popular environmental slogans such as 'Think globally, act locally'. Against this background, Heise draws attention to a need for cultural representations of what she calls 'a globalist consciousness' (2008: 4) – for images in literature, film and photography that can help us formulate and imagine a global sense of place attachment.

This chapter traces a development from a local to a global sense of place in Heaney's poetry, comparing early poems focusing on (and named after) local places in the area where Heaney grew up to later poems that illustrate a dramatically different set of concepts and ideas relating to place and place attachment. The chapter suggests that, in his later poems, Heaney looks for ways to represent the kind of global consciousness that Heise describes by reflecting a systems-oriented perspective for describing and imaging global connectedness.

The chapter draws on two important concepts introduced by Heise, to deterritorialisation and (eco-)cosmopolitanism. The two poems from *Wintering Out* that are discussed, 'Anahorish' and 'Broagh', are characterised as examples of reterritorialisation, preceding a process of deterritorialisation seen in Heaney's later poems, starting with *Station Island* (1984), especially in 'Making Strange' and continued in 'The Birthplace'. In *The Haw Lantern* (1987), the poem 'From the Land of the Unspoken' introduces the idea of imagining a global, interconnected context by picturing the world as a system. The idea of being 'at home' in a natural, non-human system is finally discussed in relation to two recent poems by Heaney; the title poem from *Electric Light* (2001) and 'A Herbal' from *Human Chain* (2010).

From 'The Blue Marble' to planetary boundaries

Heise suggests that 'a globalist consciousness has forcefully been taking shape ever since space flight enabled the first views of Planet Earth from outer space in the 1960s' (2008: 4). Photographs taken from space in the late 1960s and early 1970s effectively illustrate an idea of the earth as a single unit, with unique and limited resources. The most famous examples, 'Earthrise' (1968) and 'The Blue Marble' (1972), strongly suggested, Heise points out, the idea 'of a unified and balanced world'. These photographs inspired not only popular culture and environmental movements, but also resonated with academic scholars from fields 'as diverse as media theorist Marshall McLuhan and atmospheric scientist James Lovelock'. Heise notes the paradoxical result that images of the 'Blue Planet' resulted in 'an antitechnological rhetoric relying on an image produced by advanced technology' (2008: 20–3).[1]

One of the expressions of this new holistic view of the earth was James Lovelock's Gaia theory, first proposed in a number of articles in the 1970s and then popularised in the book *Gaia: A New Look at Life on Earth* published in 1979. Lovelock argued that the entire planet can be regarded as a single, self-regulating organism that strives to uphold optimal conditions for life on earth. Heise notes that 'the popular conception of the Gaia hypothesis became a shorthand for holistic approaches to the natural environment that emphasized balances, interdependencies, and the need for preservation rather than scientific analysis and technological exploitation' (2008: 24). In 1983, Lovelock together with Andrew Watson presented a computer model called 'Daisyworld' that illustrated a deliberately oversimplified systems model showing the biosphere as a self-regulating mechanism (Watson and Lovelock, 1983). Based on the belief that the planet itself, undisturbed by humans, would maintain optimal conditions for human and other forms of life, Lovelock argued for environmental policies based on conservation and preservation.

Ideas and theories that in different ways analyse and describe the earth as a complex integrated system have continued to prevail in scientific as well as popular imaginations in the late twentieth and early twenty-first century. This is the basic principle of the field of ecology, and in a wider sense of general systems theory, introduced as a scientific field in 1968 (by Karl Ludwig von Bertalanffy in *General System Theory: Foundations, Development, Applications*). Systems-oriented approaches designed to address social and ecological relationships have since become increasingly important. An article published in *Nature* in 2009 proposed nine different 'planetary boundaries' that it suggested could be used to identify 'a safe operating space for humanity' (Rockström et al., 2009). The article was accompanied by an illustration of the earth on which the nine planetary boundaries and their current levels are superimposed. A green area in the centre tentatively marks humanity's 'safe operating space'. The intention is that the nine different boundaries will be used as a way to assess how much humans can disturb the earth systems before risking the entire planet reaching a so-called tipping point, beyond which conditions may change dramatically. This image has been modified and reused in a number of different contexts, among others by the popular science magazine *New Scientist*, by the International Geosphere-Biosphere Programme (IGBP), and by Oxfam International, and was recently referred to as an 'iconic figure' by Helmuth Trischler, director of research at the Deutsches Museum and co-director of the Rachel Carson institute in Munich.[2]

Different from Lovelock who argued for an approach to environmental concerns based on preservation and conservation, motivated by the belief that the earth strives to maintain optimal conditions for life and that biodiversity in itself makes ecosystems more 'robust', the researchers behind the 'planetary boundaries' article argue for a more proactive approach, based on principles of governance rather than conservation, for managing and manipulating the earth's systems so that conditions of different subsystems stay within boundaries that are desirable for human societies. This might include large-scale interventions, such as geoengineering projects to regulate the climate, for instance.

The different ideas behind Lovelock's approach and the ideas behind the 'planetary boundaries' is indicated by a comparison of the related illustrations: a comparison between the figure illustrating the 'planetary boundaries' and 'The Blue Marble' photograph of the earth seen from space reflects the change in perspectives and planetary thinking that has taken place in the past fifty to sixty years, the period that O'Driscoll refers to as 'the Heaney era'. The comparison highlights the change from a view of the planet where no humans are visible, inspiring feelings of awe and wonder, to an image that is overtaken by scientific estimations and calculations of human impact and how the different systems can be manipulated to benefit future human progress and development. There is a shift in agency, from the earth's own agency as suggested by Lovelock in the 1970s, to a focus on human agency and control in the illustration of the 'planetary boundaries' from 2009.

Heise points out that, since the 1960s, environmentalist discourses have 'evolved in a field of tension between the embrace of and the resistance to global connectedness, and between the commitment to a planetary vision and the utopian reinvestment in the local' (2008: 21). The increasing number and detail of scientific representations of the global environment, such as the 'planetary boundaries' image, create a need for corresponding cultural representations, of new ways of imagining place in a global sense:

> In this multidisciplinary debate, the question of what cultural and political role attachments to different kinds of space might play, from the local and regional level all the way to the national and global, has assumed central importance. Literary and cultural critics as well as anthropologists, sociologists, historians, philosophers and political scientists have investigated the imaginative strategies and devices that allow individuals and communities to form attachments to these different types of spaces and to maintain them over time as an integral part of their identities, and have explored what overarching cultural and ideological purposes such commitments have been made to serve in different communities.
>
> (Heise, 2008: 4–5)

Many environmental movements focus on local issues, resisting different forms of globalisation. While local campaigns remain important, Heise suggests that increasing attention also needs to be directed towards the global in a positive sense.

As part of this development, Heise draws attention to alternatives to nation-based concepts of identity that have appeared in the past decades, contributing to 'an abundance of cultural studies that were skeptical vis-à-vis local rootedness and instead validated individual and collective forms of identity that define themselves in relation to a multiplicity of places and place experiences'. The idea of deterritorialisation, defined as a 'weakening of the ties between culture and place', reflects part of this process, suggesting how new forms of belonging arise that are not tied primarily to place. Globalisation theories, Heise suggests, describe new cultures that are not anchored in a specific locality, a change that ecologists have yet to incorporate into their thinking:

> Undoubtedly, deterritorialization, especially when it is imposed from outside, is sometimes accompanied by experiences of loss, deprivation, or disenfranchisement that environmentalists have rightfully resisted and should

> continue to oppose. Yet deterritorialization also implies possibilities for new cultural encounters and a broadening of horizons that environmentalists as well as other politically progressive movements have welcomed, sometimes without fully acknowledging the entanglements of such cultural unfolding with globalization processes that they otherwise reject. The challenge that deterritorialization poses for the environmental imagination, therefore, is to envision how ecologically based advocacy on behalf of the nonhuman world as well as on behalf of greater socioenvironmental justice might be formulated in terms that are premised no longer primarily on ties to local places but on ties to territories and systems that are understood to encompass the planet as a whole.
>
> (2008: 5)

To think about a global environment and sense of belonging, Heise combines the idea of deterritorialisation with theories describing cosmopolitanism and globalisation. Heise specifies the concept of cosmopolitanism as a kind of 'eco-cosmopolitanism', referring to an 'environmental world citizenship, building on recuperations of the cosmopolitan project in other areas of cultural theory' (2008: 10). In the late 1990s, theorists from different fields began to use the term cosmopolitanism in order to express attachments to places not defined by localism or nationality. Cosmopolitanism in this sense can be seen as describing a positive outcome of deterritorialisation; a sense of global place attachment replacing a weakened local equivalent. Heise refers to several theories that attempted to give cosmopolitanism a new significance, different from its earlier associations with 'social privilege and leisure travel':

> While there are considerable differences in the way these theorists rethink cosmopolitanism, they share with earlier theorists of hybridity and diaspora the assumption that there is nothing natural or self-evident about attachments to the nation, which are on the contrary established, legitimized, and maintained by complex cultural practices and institutions. But rather than seeking the grounds of resistance to nationalism's nation-based identities in local communities or groups whose mobility places them at the borders of national identity, these theorists strive to model forms of cultural imagination and understanding that reach beyond the nation and around the globe.
>
> (2008: 6)

After a wave of 'countercritiques' that argued for the importance of local and national belonging as foundations for identity and resistance in a globalised world, the debate between local and global place attachment has according to Heise now reached 'a conceptual impasse: while some theorists criticize nationally based forms of identity and hold out cosmopolitan identifications as a plausible and politically preferable alternative, other scholars emphasize the importance of holding on to national and local modes of belonging as a way of resisting the imperialism of some forms of globalization' (2008: 7). The result is of two complementary or opposing theoretical viewpoints, where local and national identities are seen as essentialist and oppressive on the one hand, and as forms of resistance to global domination on the other. Heise quotes Arif Dirlik who has commented on this theoretical stalemate that 'the defense and the repudiation of place both carry considerable theoretical plausibility and for that same reason seem in their opposition to be confined within a theoretical world of their own out of which there is no exit that is to be revealed by theory', concluding that in order to move forward 'the entire discussion should be shifted to the level of specific case studies' (quoted by Heise, 2008: 7–8). This chapter attempts such a case study of Heaney's poetry. In Heaney's poetics, there is a growing tension between on the one hand advocacy and strengthening of local ties, cultures and belonging, and on the other hand a growing recognition of global connectedness, with regard to environmental and other relationships. This change from local to global, while relating to abstract developments and concepts on some levels, also depends on the particularity of Heaney's local context, in particular the troubled past of Ireland and Northern Ireland.

Sense of place

Heaney discusses notions of place and place attachment in a lecture from 1977 titled precisely 'The Sense of Place', included in *Preoccupations*. Heaney introduces the lecture by stating that:

> I think there are two ways in which place is known and cherished, two ways which may be complementary but which are just as likely to be antipathetic. One is lived, illiterate and unconscious, the other learned, literate and conscious.

The primary topic of the lecture is the meaning of place in the specific context of Ireland and Northern Ireland, but some observations are applicable to a more general discussion of place and place attachment. Heaney notes the difference between places that have a widespread cultural meaning and more anonymous places where a typical visitor has no preconceptions that provide the place with a specific meaning or framework. In some cases, the lack of cultural meaning is not due to an absence of tradition, but rather to the fact that older traditions and cultures are no longer part of the common or popular imagination:

> When we go as tourists to Donegal or Connemara or Kerry we go with at best an aesthetic eye, comforting ourselves with the picturesqueness of it all or rejoicing in the fact that it is unspoiled. We will have little felt knowledge of the place, little enough of a sense of wonder or a sense of tradition. Tory Island, Knocknarea, Slieve Patrick, all of them deeply steeped in associations from the older culture, will not stir us beyond a visual pleasure unless that culture means something to us, unless the features of the landscape are a

> mode of communion with a something other than themselves, a something to which we ourselves still feel we might belong.
>
> (Heaney, 1980: 131)

Heaney compares these kinds of anonymous places with other places that 'now live in the imagination', that 'stir us to responses other than the merely visual'. When confronted with such places, Heaney suggests,

> our imaginations assent to the stimulus of the names, our sense of the place is enhanced, our sense of ourselves as inhabitants not just of a geographical country but of a country of the mind, whether that country of the mind takes its tone unconsciously from a shared oral inherited culture, or from a consciously savoured literary culture, or from both, it is this marriage that constitutes the sense of place in its richest possible manifestation.
>
> (1980: 132)

This reflects the process undertaken in Heaney's bog poems, as suggested by Dianne Meredith's analysis of the transformation of the 'bogscape' into a 'mindscape' discussed in Chapter 3 in this book.

As another example, Heaney refers to the poem 'Inniskeen Road, July Evening' by Patrick Kavanagh, and reads the dialectical expressions in that poem as ways to inscribe it within the tradition and culture of the particular place it refers to. Rather than describing an outsider's view of the place, Heaney states that the use of dialect signals that Kavanagh 'meets his people at eye-level, he hears them shouting through the hedge and not through the chinks in a loft floor, the way Synge heard his literary speech in Co. Wicklow' (1980: 138).

However, Heaney also states that, compared to a poet like John Montague, for instance, Kavanagh is not really a place-based poet:

> Kavanagh's sense of his place involves detachment, Montague's attachment. When Montague asks who he is, he is forced to seek a connection with a history and a heritage; before he affirms a personal identity, he posits a national identity, and his region and his community provide a lifeline to it. Whereas Kavanagh flees the abstractions of nationalism, political or cultural. To find himself, he detaches rather than attaches himself to the communal. I rather than we is his preferred first person.
>
> (1980: 143–4)

The difference could be described as a difference between local and national, or between local and global. While the local is a directly experienced relation, the national or global depends on a constructed or imagined sense of identity. While formulated differently in different places, there is a sense in which the local can also be referred to as individual, and at the same time universal: Heaney quotes Kavanagh, who states that 'Parochialism is universal; it deals with the fundamentals' (Heaney, 1980: 139). Heaney's description of Kavanagh's detachment from place compared to Montague's attachment is based on this difference between local/individual/universal and regional/national/tribal: 'Kavanagh's place names are there to stake out a personal landscape, they declare one man's experience, they are denuded of tribal or etymological implications', while Montague's place names 'are rather sounding lines, rods to plumb the depths of a shared and diminished culture [...] redolent not just of his personal life but of the history of his people, disinherited and dispossessed' (Heaney, 1980: 140, 141).

Heaney concludes 'The Sense of Place' by looking forward and stating that in modern, mobile times 'We are no longer innocent, we are no longer just parishioners of the local.' He adds that, despite a loss of innocence, he remains convinced that 'those primary laws of our nature are still operative' and that it is to 'the stable element, the land itself, that we must look for continuity' (1980: 148–9).

'Anahorish' and 'Broagh'

In a study from 2002, Eugene O'Brien argues that Heaney moves from a position close to Montague's towards one closer to Kavanagh's: from place attachment to detachment, from drawing on the communal and regional to focusing on the (at the same time) individual and universal (O'Brien, 2002a). The poems 'Anahorish' (Heaney, 1972: 16) and 'Broagh', both from *Wintering Out*, illustrate Heaney's early sense of place attachment. Anahorish is the name of Heaney's primary school, and the poem describes the speaker's early experiences of this place:

> My 'place of clear water',
> the first hill in the world
> where springs washed into
> the shiny grass
>
> and darkened cobbles
> in the bed of the lane.

Focusing on the word 'Anahorish', the next stanza states:

> *Anahorish*, soft gradient
> of consonant, vowel-meadow,
>
> after-image of lamps
> swung through the yards
> on winter evenings.

Greg Garrard notes that in this poem 'naming, thoughtfully carried out, shows that the poet is at "home"', and suggests that Heaney's poetry also in a more general sense exhibits 'a naming of places which is at once (and perhaps paradoxically) restorative and commemorative' (1999: 186). He describes the phrase '*Anahorish*, soft gradient / of consonant, vowel-meadow' as 'poetic topology' (1999: 187), which, a few lines further on, turns into something he calls 'mythotopology':

> Heaney moves suddenly from a near-contemporary 'after-image of lamps' to the pre-historic 'mound-dwellers' in a single stanza break, registering the archaic watery resonances in *Anahorish*. We might call this something like 'mythotopology' because of the sense of unbroken continuity developed between the poet and the ancient keepers of the 'first hill in the world'.
>
> (1999: 188)

The idea of 'mythotopology' reflects the meaning of place discussed by Heaney above as derived from older, often forgotten, cultures and traditions. It is between ancient history in this sense and the poet that an 'unbroken continuity' is established in 'Anahorish'.

The place described in 'Anahorish' is, however, an idealised place. Garrard notes that the '"ecolectal" voice' in the poem 'is possible in this unproblematic way only because all relevant historical differences – between Catholic and Protestant, Gaelic and Hiberno-English, present-day and Bronze Age inhabitants – have been suppressed' (1999: 189). When realised, this idealisation of history undermines the attachment between poet and place in the poem, which then rather suggests an imaginary reconstruction of an idealised past. The idealisation of place in 'Anahorish' is qualified in a later poem in *District and Circle*, 'Anahorish 1944' (Heaney, 2006: 7). This poem depicts the same place in different terms, describing the arrival of American soldiers in Anahorish:

> Two lines of them, guns on their shoulders, marching.
> Armoured cars and tanks and open jeeps.
> Sunburnt hands and arms. Unknown, unnamed,
> Hosting for Normandy.

Like the different poems about the Tollund man, this poem creates a sense of both change and continuity between poems from different points in Heaney's poetics. Reading this later poem about 'Anahorish' also makes the idealisation that takes place in the first visible.

The poem 'Broagh' also describes a place from Heaney's childhood. Like the 'vowel-meadow' in 'Anahorish', Garrard notes that in this poem the word Broagh is also described as part of the physical environment, almost as if it was pronounced by the landscape itself:

> The garden mould
> bruised easily, the shower
> gathering in your heelmark
> was the black O

in *Broagh*

The link that is established in these lines between the physical place and its name suggests a sense of continuity between the two different experiences of place that Heaney refers to in the beginning of 'The Sense of Place', between a 'lived, illiterate and unconscious' experience on the one hand, and a 'learned, literate and conscious' experience on the other. This link is developed in the next lines:

> *Broagh*,
> its low tattoo
> among the windy boortrees
> and rhubarb-blades
>
> ended almost
> suddenly, like that last
> gh the strangers found
> difficult to manage.

The last two stanzas suggest a difference between 'us' and 'them', comparing a native's experience of the place to an outsider's. Garrard notes that this difference 'introduces an element of discord into the tidy picture created by the mythotopology' (1999: 189). This mythotopology reflects the relationship between nature and culture in a particular place, captured and concentrated in this poem in the sound and pronunciation of the word 'Broagh'.

Like 'Anahorish', 'Broagh' also establishes and confirms a relationship between speaker and place; as noted by O'Brien, both poems signify 'a politicisation of the land' (2002b: 51), reflecting, as also noted by David Lloyd, a relationship between poet and place as indicating a 'foreclosed surety of the subject's relation to place, mediated as it is by a language which seeks to naturalize its appropriative function' (O'Brien, 2002b: 52).

Based on the affirmative descriptions of the relationship between poet and place, 'Anahorish' and 'Broagh' are poems of re- rather than deterritorialisation as outlined by Heise. Heaney's later collections suggest a move away from this kind of affirmative descriptions of place attachment towards questioning the idea of rootedness and the relationship between place and identity. O'Brien argues that Heaney moves away from speaking from a tribal and communal point of view (similar to Montague, in Heaney's description) towards developing an identity politics that is self-centred rather than communal (more like Kavanagh's). This move can be interpreted as a move towards an increasingly flexible sense of place and belonging, formulated more in relation to inner than outer nature. A move in this direction, indicating the beginning of a process of deterritorialisation in Heaney's poetics, is found in two poems from *Station Island*: 'Making Strange' and 'The Birthplace'.

'Making Strange' and 'The Birthplace'

Referring to *North*, Richard Kirkland suggests that 'Beyond representation, the poetry becomes an embodiment of the real angst of the community and its place, while Heaney is absolutely assimilated into its place' (1997: 260). After *North*, however, the poems begin to open up towards the outside world, to new perspectives and influences. Part of this process involves a focus on the individual as representative of the universal, rather than on the communal, regional or national.

'Making Strange' (Heaney, 1984: 32–3) illustrates an outsider's perspective on Northern Ireland. Kirkland suggests that this poem 'subtends its nominal consideration of identity as expressed in relation to geography under a desire to achieve a satisfactory form of coherent closure within its own formal limits' (1997: 263). Different from 'Anahorish' and 'Broagh', 'Making Strange' describes an area from the poet's childhood not through the poet's own memories, but through the eyes of a stranger. The poem juxtaposes the outsider's view of this place with the perspective of one of the locals, with the poet himself caught between the two:

> I stood between them,
> the one with his travelled intelligence
> and tawny containment,
> his speech like the twang of a bowstring,
>
> and another, unshorn and bewildered
> in the tubs of his wellingtons,
> smiling at me for help,
> faced with this stranger I'd brought him.

Kirkland refers to Michael Parker's explanation 'that this poem has its genesis in a guided tour of South Derry undertaken by Heaney for the benefit of the Jamaican poet Louis Simpson' (Kirkland, 1997: 263). Heaney describes the origin of the poem in an interview:

> It started from a chance meeting between my father, myself and Louis. He was being driven by me from Belfast to a poetry reading in the University of Ulster at Coleraine, so I took a detour through my part of the country and stopped at the pub next to our old Mossbawn house. Next thing, my father appears on the scene, and is being included as more or less part of the tour. So there I was between the pair of them, at home and not at home, between the Toome Road and the open road.
>
> (O'Driscoll, 2008: 113)

The poem continues by introducing a third voice onto the scene: 'Then a cunning middle voice / came out of the field across the road'. The voice encourages the speaker to pay attention to his local surroundings. As discussed in Chapter 4 in relation to ecological and historical and political linguistics, this voice also draws attention to the relationship between the natural environment and its vernacular names:

> Be adept and be dialect,
> tell of this wind coming past the zinc hut,
>
> call me sweetbriar after the rain
> or snowberries cooled in the fog.

However, the voice then continues by telling the speaker to also 'love the cut of this travelled one / and call me also the cornfield of Boaz'. Paul Keen suggests that this voice is the voice of poetry, indicating that the solution to the conflicting views of the visitor and Heaney's father is not a separation of notions but rather a 'dialectical engagement – poetry as a place where the conversations between these different voices can be heard' (1996: 77). It could also be interpreted as the voice of the place itself. In this sense, it ties in with references to dialect in other poems by Heaney, as the voice of place is also dialect. In this reading, the physical landscape is given a voice of its own, contrasting the perspectives of the same place heard through the voices of Heaney's father and of the stranger. Just like the voice of poetry could be a place for 'dialectical engagement' as suggested by Keen, so could place provide a frame of reference for conversations between different voices.

The combination of 'adept' and 'dialect' is clarified by the encouragement of the strange voice to 'love the cut of this travelled one' while at the same time 'call me also the cornfield of Boaz'. The cornfield of Boaz refers to Keats's 'Ode to a Nightingale', where the character Ruth is 'sick for home', 'stood in tears amid the alien corn' (II. 66–7). The juxtaposition is between the strange and the familiar, of staying close to home and moving away, suggesting a conflict of interest between local and global that the speaker attempts to reconcile. In Keats's poem the speaker also hears a voice, which he acknowledges was also 'heard / In ancient days by emperor and clown' (II. 63–4), suggesting that the 'cunning middle voice' in Heaney's poem is not unique, but universal. The poem might suggest that the need for appropriate forms that can accommodate both sides of the juxtaposition can be achieved either through the means of poetry, as Keen suggests, or through an appropriate notion of place that manages to be at the same time local and global.

The poem concludes with the poet proceeding to show the stranger the area, following the advice of the 'middle voice':

> I found myself driving the stranger
>
> through my own country, adept
> at dialect, reciting my pride
> in all that I knew, that began to make strange
> at that same recitation.

The defamiliarisation process described in these lines, an effect of the poet seeing the area through the eyes of the outsider, can also be described as a deterritorialisation process that nuances or challenges the speaker's existing view of this place. The poem indicates how the introduction of the outside world (in the form of the visitor) into the speaker's home territory makes the familiar landscape appear strange, weakening the kind of place attachment that is affirmed in poems such as 'Anahorish' and 'Broagh'. Whereas in 'Broagh' at the end of the poem outsiders' inability to correctly pronounce the name of the place was used to identify 'us' versus 'them', in 'Making Strange' the outsider's point of view is accommodated and allowed to make the familiar place take on an at least partly new meaning through an unfamiliar, cosmopolitan perspective. In this way, 'Making Strange' may point to Heise's description of deterritorialisation as indicating 'how experiences of place change under the influence of modernization and globalization processes' (Heise, 2008: 51).

Deterritorialisation is part of an alternative to a sense of belonging primarily anchored in the local. Heise suggests that, while 'affirmations of local ties can play an important role in environmentalist struggles', environmentalists also need 'to come to terms with one of the central insights of current theories of globalization: namely, that the increasing connectedness of societies around the globe entails new forms of culture that are no longer anchored in place' (Heise, 2008: 10). Heaney's poem 'The Birthplace' (about Thomas Hardy, Heaney, 1984: 34–5), also from *Station Island*, suggests a deterritorialisation process from a different perspective than 'Making Strange'. Its third and final part begins by asking:

Everywhere being nowhere,
who can prove
one place more than another?

The following lines further denounce a local sense of place, as well as feelings of being at home and at rest. Instead the poem describes a sense of restlessness, of being 'afloat among galaxies':

We come back emptied,
to nourish and resist
the words of coming to rest:

birthplace, roofbeam, whitewash,
flagstone, hearth,
like unstacked iron weights

afloat among galaxies.

The 'unstacked iron weights' suggest feelings of relief rather than distress, perhaps to be read in relation to Heaney's personal experience of being weighed down by a feeling of specific artistic responsibilities as a Northern Irish poet.

The sense of weightlessness is more ambiguous in the description of being 'afloat among galaxies', reflecting the dual notion of 'nourish and resist'.

The feelings of weightlessness in these lines are undermined by a recognition of a persisting gravitation towards a particular place in the remaining lines of the poem where, as in 'Broagh', written or spoken words merge with the speaker's immediate, physical surroundings:

> Still, was it thirty years ago
> I read until first light
>
> for the first time, to finish
> *The Return of the Native?*
> The corncrake in the aftergrass
>
> verified himself, and I heard
> roosters and dogs, the very same
> as if he had written them.

Despite these final lines, O'Brien describes 'The Birthplace' as a 'liberating comment on place, that concept which was so central to the earlier books, where it was shot through with connotations of racial and communal identity and territory'. Lines like those in the beginning of this poem's third part, about 'Everywhere being nowhere', are described by O'Brien as 'emblematic of the process at work in this section of Heaney's work, as he takes cultural, linguistic and historical givens, and attempts to transcend them through his writing' (2002a: 61–2). This idea is developed further in Heaney's next collection, *The Haw Lantern* (1987).

'From the Land of the Unspoken'

In one of the lines of the poem 'From the Land of the Unspoken' (Heaney, 1987: 18–19), from *The Haw Lantern*, the speaker declares that 'I could feel at home inside that metal core / slumbering at the very hub of systems'. Commenting on this poem, George Morgan asks Heaney in an interview from 1998 what the notion of 'home' or being 'at home' means to him. Heaney replies:

> It's not something I've thought through. I try to dive down, as far as possible, away from the analyses and the self-consciousness encouraged by these questions. I'm not quite sure what 'home' means other than that deep sense of planetary, experiential creaturely, animal 'at-homeness' which I'm trying to express in the metaphor of 'slumbering at the heart of systems'. When I hear the word 'home', I hear the sound the earth might make humming on its axis. Something Wordsworthian. You know that poem 'A slumber did my spirit seal' where Wordsworth says that Lucy is 'roll'd round in earth's diurnal course / With rocks and stones and fields'. 'At home' means something like that to me.

Heaney explains this feeling of 'at-homeness' further as 'Something centred and deeply harmonious':

> I'd have to be a bit Heideggerian here, and coin a term like 'in-dwelling'. My favourite meditation on it is in a poem called 'The Birthplace' which is about Thomas Hardy's birthplace. But I'd want to insist that when I use the word 'home', I'm not talking of 'hominess' in the American sense. I credit all that nostalgic part of my being but I also want to put it to the test, to remind it that we are made up of homelessness as well as centredness.
>
> (Morgan, 1998)

This reflection on nostalgia versus homelessness perhaps alludes to nostalgic and idealising notions in the early poems, like 'Boragh' and 'Anahorish', counteracted in later poems that reflect on more complex and conflicting notions of being 'at home'.

The first stanza of 'From the Land of the Unspoken', where the description of 'the hub of systems' appears, reads:

> I have heard of a bar of platinum
> kept by a logical and talkative nation
> as their standard of measurement,
> the throne room and the burial chamber
> of every calculation and prediction.
> I could feel at home inside that metal core
> slumbering at the very hub of systems.

The 'bar of platinum' seems to refer to the international prototype metre, kept by the International Bureau of Weights and Measurements in Sèvres, France. This prototype was created in 1889 and kept as a standard unit for the metric system until 1960 (when a new standard was created using wavelength measurements). The platinum prototype represents something standardised and universal, with no room for local particularities or dialects. The reference of being at home in this measurement, which is exactly the same everywhere, points to the notion of one place being like all others, as suggested in 'The Birthplace'. In 'From the Land of the Unspoken' the speaker experiences a feeling of being primarily at home not in a local context, or in any locality at all, but rather in the very fact that all places belong to the same system, and are defined by the same rules. In this sense, the speaker is at home in the world or planet as a whole, as a place defined by those rules. This is perhaps what Heaney refers to in the interview when he relates the notion of being at home 'at the heart of systems'[3] to a sense of planetary 'at-home-ness'.

The 'standard of measurement' is to a description of the world that is based on laws of nature rather than on human language, which also explains the title of the poem, 'From the Land of the Unspoken', referring to a land governed by laws that are not made by humans, that are mathematical rather than lingual. The poem's second stanza describes humanity's 'exile' from this world of the 'unspoken':

We are a dispersed people whose history
is a sensation of opaque fidelity.
When or why our exile began
among the speech-ridden, we cannot tell

In this poem, the notion of being at home in a particular place and in a certain dialect is replaced by the idea of returning to being at home in a system defined by non-human laws and languages. The feeling of being 'at home inside that metal core / slumbering at the very hub of systems' suggests a sense of planetary belonging based on the fact that the 'metal core' represents a fundamental natural principle that is present everywhere and that makes each place similar to all other places, as suggested in 'The Birthplace'. The system that the platinum bar indicates or symbolises relates to the systems governing the planet that the figure illustrating the 'planetary boundaries' tries to capture. In a sense similar to that illustration, 'From the Land of the Unspoken' can be read as suggesting a systems-oriented perspective for understanding the interconnectedness of the different parts of and places on the planet. Being at home in this non-human system also provides a sense of connectedness with non-human nature. However, the poem also recognises that, while natural laws are the same everywhere, there are significant differences, as in the line 'When we recognize our own, we fall in step / but do not altogether come up level'. O'Brien notes that the ambiguity of place and place attachment suggested in this poem and others from the same period counteracts 'those appetites of gravity' (2002a: 105) that animated earlier collections, especially *North*.

'Electric Light' and 'A Herbal'

A different system is described in 'Electric Light', from the collection with the same name. The arrival of electric light enabled new perspectives on the familiar, suggesting a defamiliarisation process similar to that described in 'Making Strange'. O'Brien notes that:

> Electric light allows us to see in the dark, to see where we could not see before, to see things anew. *Electric Light* symbolises such a new perspective, as personal, cultural and political events are seen through the alembic of other cultures, literatures and languages in such a way as to see them anew.

O'Brien also points out that some of the images in 'Electric Light' are referred to in Heaney's Nobel lecture from 1995, especially the image of the radio. He suggests that 'the actual image from "Electric Light" that is most significant is the radio dial, which allows Heaney to experience the different languages and cultures of the world through the radio "stations of the world"' (2002a: 169). In his Nobel lecture, Heaney describes how the radio brought the outside world into the rural kitchen of his childhood:

> When a wind stirred in the beeches, it also stirred an aerial wire attached to the topmost branch of the chestnut tree. Down it swept, in through a hole bored in the corner of the kitchen window, right on into the innards of our wireless set where a little pandemonium of burbles and squeaks would suddenly give way to the voice of a BBC news reader speaking out of the unexpected like a *deus ex machina.*

As Heaney grows older, his listening becomes 'more deliberate':

> Now that the other children were older and there was so much going on in the kitchen, I had to get close to the actual radio set in order to concentrate my hearing, and in that intent proximity to the dial I grew familiar with the names of foreign stations, with Leipzig and Oslo and Stuttgart and Warsaw and, of course, with Stockholm.

This introduction of the foreign into the familiar, especially in the form of foreign languages, turned Heaney's imagination away from his local surroundings and outwards to the unknown:

> I also got used to hearing short bursts of foreign languages as the dial hand swept round from BBC to Radio Eireann, from the intonations of London to those of Dublin, and even though I did not understand what was being said in those first encounters with the gutturals and sibilants of European speech, I had already begun a journey into the wideness of the world beyond.
>
> (Heaney, 1995)

Heaney relates this outer change to a corresponding inner change, from being 'emotionally and intellectually proofed against the outside world' to gradually gaining a sense of the unfamiliar, through language in general and through poetry especially. This change is reflected in the title of the Nobel lecture, 'Crediting Poetry'. Poets like Kavanagh, Bishop and Lowell gave Heaney what he describes as 'reasons for believing in poetry's ability – and responsibility – to say what happens, to "pity the planet", to be "not concerned with Poetry"'.

'Electric Light' (Heaney, 2001: 80–1) describes the arrival of electric light in the area where Heaney grew up, and contrasts images of the new invention with images of tradition, of things that remain the same 'Year in, year out':

> In the first house where I saw electric light,
> She sat with her fur-lined felt slippers unzipped,
>
> Year in, year out, in the same chair, and whispered
> In a voice that at its loudest did nothing else
> But whisper.

The speaker remembers that they 'were both desperate', and how he 'wept and wept' under a 'waste of light'. The poem contrasts the old woman, using the word '*Ails*, far off and old' with the new, intruding light. A few lines further on, the speaker recalls how he first operated the radio dial, and how the older people 'watched me / As I roamed at will the stations of the world'. The poem describes how the distant and global are introduced into the local and familiar:

> If I stood on the bow-backed chair, I could reach
> The light switch. They let me and they watched me.
> A touch of the little pip would work the magic.
>
> A turn of their wireless knob and light came on
> In the dial. They let me and they watched me
> As I roamed at will the stations of the world.

O'Brien suggests that the sense of attachment to the image of the radio in 'Electric Light' signifies a process of deterritorialisation together with a sense of cosmopolitanism, suggesting

> an Irishness that is centrifugal as opposed to centripetal in orientation. Here we see an embracing of European and world culture, an unselfconscious placement of Irish experience in the context of such a culture, and a willingness to posit connections between the two. Through electricity, the light of different cultures and languages, the 'stations of the world', came into the home and mind of Seamus Heaney, and this is celebrated in the cosmopolitan, sophisticated and nuanced sense of Irishness, as well as in the complexity of identity that is enunciated throughout his writing.
>
> (O'Brien, 2002a: 170)

The electrical grid that introduces the outside world into Heaney's childhood can be compared to that of the metric system described in 'From the Land of the Unspoken'. Both systems signify a connectedness between the local to the global, relating that which is immediately experienced to a system that is of a different order and magnitude. In this sense the idea of a global interconnectedness replaces, or at least complements, the local.

However, both the metric system and the electrical grid also manifest themselves locally in artefacts or results, such as a lamp or a measurement. By thus translating a global or planetary system into local and immediate experience, Heaney creates a sense of continuity between the local and the global. Moving between these different perspectives and contexts, *Electric Light* creates an intersection of what O'Brien calls on the one hand a 'cosmopolitan range of names and places', and on the other 'a number of local names and places which take their place in this constellation' so that 'What the book achieves is the placement of these different cultures in the same structure' (2002a: 162–3).

O'Brien argues that, from *Station Island* and onwards, Heaney 'is looking for plural sources of selfhood, and for a more fluid and distanced relationship with place'. By the end of *Electric Light*, his 'relationship with place has been transformed, from an artesian probing of the psychic communal memory bank to a more individualistic translating and transforming of that past into a personalised aesthetic wherein the relationship with tradition is more nuanced and the perspective is more transcendental than immanent' (O'Brien, 2002a: 69). In the sense that Heaney describes with reference to Kavanagh (in contrast to Montague), this transformation of communal into individual also signifies a move from the communal to the universal. Kirkland notes that, based on his move from local to cosmopolitan, 'Heaney is the physical embodiment of George Moore's belief that art "must be parochial in the beginning to become cosmopolitan in the end"' (1997: 254).

O'Brien also notes that this new sense of place represented by Heaney reflects how 'the sense of place and identity exists in the mind as opposed to in the land itself' (2002a: 90). A similar idea is captured in the poem 'A Herbal' (Heaney, 2010: 35–43), from *Human Chain*, which describes the relationship to place as 'Me in place and the place in me'. At the end of this long poem, the speaker states:

> Between heather and marigold,
> Between sphagnum and buttercup,
> Between dandelion and broom,
> Between forget-me-not and honey-suckle,
>
> As between clear blue and cloud,
> Between haystack and sunset sky,
> Between oak tree and slated roof,
>
> I had my existence. I was there.
> Me in place and the place in me.

These lines connect the speaker to the surrounding natural environment, but acknowledge at the same time that his sense of this place comes as much from within as from without. The next lines ask how the physical experience of place can be translated into an experience of the world beyond the immediate senses; how to combine the local and sensory with the planetary and abstract:

> Where can it be found again,
> An elsewhere world, beyond
>
> Maps and atlases,
> Where all is woven into
>
> And of itself, like a nest
> Of crosshatched grass blades?

These lines suggest both the idea of a global interconnectedness and the world as a system, in the reference to 'Maps and atlases', but also ask how that sense of planetary context and connectedness can be represented in a way that relates it to the experience of the local environment, signified by the 'crosshatched grass blades'.

Perhaps Heaney accomplishes this connection through what Irene Nordin calls 'a poetry of more detachment and wider magnitude', which she sees in Heaney's collections from *The Haw Lantern* and onwards. Nordin argues that Heaney's early work

> is very much a poetry of the self and of the Irish rural background in which he was born and in which he grew up. There is a deep element of what he himself calls 'the sense of place' – the sense of being a part of the very landscape, both geographical and historical. [...] personal memory and a sense of the local have been vital ingredients in his poetry from the start.
> (2000: 177)

Like O'Brien and Kirkland, Nordin argues that there is a development in Heaney's poetry from a parochial focus in the early poems in the collections from the 1960s, to a more distant perspective in the later poems, where farming memories and the rural tradition of Heaney's own childhood is looked at in relation to a consciousness of a different, global and cosmopolitan belonging and culture. Nordin suggests that 'In this way the personal preoccupations and the self-conscious experiences of his earlier poetry are reimagined and transformed and seen in a wider, more encompassing, focus' (2000: 174). This development from the private to the universal is also a redefinition of the personal – from a sense of the personal defined in relation to communal identity and a local sense of belonging to a personal identity understood in relation to the universal. The transformation is thus, as Kirkland also notes, less from the personal to the universal than from the local and regional to the universal.

This transformation affects the understanding of local landscapes. Rather than a direct understanding of what is perceived or cultivated, such as the farming practices of 'Digging' for example, in Heaney's later poems landscapes are also understood in relation to their invisible relations to places elsewhere. Nordin describes the landscape in Heaney's later poetry as 'sacramental, instinct with signs, implying a system of reality beyond the visible realities' (2000: 174), a landscape such as that in *North* which relates Northern Ireland to Denmark, for example. From *The Haw Lantern* and onwards, Nordin suggests, Heaney's poetry becomes characterised by 'an artistic distance', signifying 'a visible moving away from the earlier digging "down and down for the good turf" to a poetry of more detachment and wider magnitude, where the poet's "inheritance" no longer is seen as a burden that weighs him down', so that it becomes 'a poetry of more all-embracing immanence where the looking downwards and inwards of the early work broadens to include a looking upwards and outwards as well'. The personal and local remain important, but are now understood as part of something deeper, broader, systematic and global: 'The poetic utterance thus moves beyond the personal to a deeper awareness, as the poet/speaker becomes part of a wider more universal picture, where he finally loses himself in his creation' (Nordin, 2000: 177, 178).

This increasing detachment from a limited local perspective to include an awareness of the relationship between local and global transforms the focus of Heaney's early poems from a sense-based understanding of a local environment to an understanding of a transformed relation between human, local and global natures. Sense of place and belonging in these later poems include invisible environments that the speaker only knows in an abstract, theoretical sense, reflecting and constituting a sense of planetary belonging. By shifting between personal and universal Heaney is able to create a sense of zooming in and out between the local and the global, in the sense discussed by Heise, and forge a sense of continuity between place-based experiences and planetary imageries.

Notes

1 For example, that the 'Blue Marble' photograph is used to illustrate the Wikipedia page for 'ecology' [accessed Mar. 2014].

2 At the Science and Technology Studies conference at KTH Royal Institute of Technology, in relation to the panel 'Museums in the Anthropocene: Climate Change and Social History', Stockholm, 4 May 2012.

3 Heaney seems to misquote his own poem in the interview, using the phrase 'heart of systems' instead of 'hub of systems'.

References

Garrard, G. (1999) 'Ecological Literary Criticism: Two Case Studies', doctoral dissertation, University of Liverpool.

Heaney, S. (1972) *Wintering Out*, London: Faber & Faber.

Heaney, S. (1980) *Preoccupations: Selected Prose, 1968–1978*, New York: Farrar, Straus & Giroux (first published by Faber & Faber).

Heaney, S. (1984) *Station Island*, London: Faber & Faber.

Heaney, S. (1987) *The Haw Lantern*, London: Faber & Faber.

Heaney, S. (1995) 'Crediting Poetry', Nobel lecture. http://nobelprize.org/nobel_prizes/literature/laureates/1995/heaney-lecture.html [accessed Mar. 2014].

Heaney, S. (2001) *Electric Light*, London: Faber & Faber.

Heaney, S. (2006) *District and Circle*, London: Faber & Faber.

Heise, U. (2008) *Sense of Place and Sense of Planet: The Environmental Imagination of the Global*, New York: Oxford University Press.

Keen, P. (1996) '"Making Strange": Conversations with the Irish M/Other', *Irish University Review*, 26/1: 75–87.

Kirkland, R. (1997) 'Paradigms of Possibility: Seamus Heaney', in M. Allen (ed.), *Seamus Heaney*, pp. 252–265, London: Macmillan Press.

Lovelock, J.E. (1979) *Gaia: A New Look at Life on Earth*, Oxford: Oxford University Press.

Morgan, G. (1998) 'Interview with Seamus Heaney', *Cycnos*, 15/2. http://revel.unice.fr/cycnos/index.html?id=1594 [accessed Mar. 2014].

Nordin, I.G. (2000) 'Seamus Heaney: From the Personal to the Universal', *Studio Neophilologica*, 72/2: 174–180.

O'Brien, E. (2002a) *Seamus Heaney: Creating Irelands of the Mind*, Dublin: Liffey Press.

O'Brien, E. (2002b). *Seamus Heaney and the Place of Writing*, Gainesville, FL: University Press of Florida.

O'Driscoll, D. (2008) *Stepping Stones: Interviews with Seamus Heaney*, London: Faber & Faber.

O'Driscoll, D. (2009) 'Finding the Right Words in Which to Speak for us', *Irish Times*, Apr. www.irishtimes.com/indepth/seamus-heaney/finding-the-right-words-in-which-to-speak-for-us.html [accessed Apr. 2012].

Rockström, J., Steffen, W., Noone, K., Persson, Å., Chapin, S.C. III, Lambin, E.F., Lenton, T.M., Scheffer, M., Folke, C., Schellnhuber, H.J., Nykvist, B., de Wit, C.A., Hughes, T., van der Leeuw, S., Rodhe, H., Sörlin, S., Snyder, P.K., Costanza, R., Svedin, U., Falkenmark, M., Karlberg, L., Corell, R.W., Fabry, V.J., Hansen, J., Walker, B., Liverman, D., Richardson, K., Crutzen, P., Foley, J.A. (2009) 'A Safe Operating Space for Humanity', *Nature*, 461: 472–5.

Watson, A.J. and Lovelock, J.E. (1983) 'Biological Homeostasis of the Global Environment: The Parable of Daisyworld', *Tellus B* (International Meteorological Institute), 35/4: 286–9.

Conclusion

Hughes, Heaney and the different natures of ecopoetics

This concluding section draws on the previous chapters to discuss ideas about nature, culture and environment in Hughes's and Heaney's respective ecopoetics from a broader perspective, in comparison to each other as well as in relation to a wider framework of the nature and definition of ecopoetics.[1] The first part outlines Hughes's anti-anthropocentrism, expressed through references to evolutionary and other natural processes that take place on scales beyond the apprehension of the individual human being. The second part discusses the relation in Heaney's poetry between language, ecology and his own poetic process. The final section contextualises Hughes's and Heaney's respective themes and ideas in relation to a broader discussion about the relationship between nature, environment and poetry.

Hughes's anti-anthropocentrism and nature on a non-human scale

Hughes's poems reach beyond the realm of the human in their attempts to reflect and describe nature from a non-anthropocentric perspective. Hughes's anti-anthropocentrism can be recognised in at least three different forms. The first of these has to do with scale. Frequently, Hughes's poems refer to natural processes that take place on a scale beyond a human sense of time or space. This is seen in the *Elmet* poems, for example, which, as discussed in Chapter 3, portray the passing of the last Ice Age and the subsequent formation of the moors and the Calder valley. The *Elmet* poems rely on this recognition of natural processes beyond human influence to suggest that the negative environmental impact of industrialisation on the Calder valley will eventually be erased from the area, overcome by non-human nature.

Natural forces of similarly non-human scale or influence form one of the basic themes of Hughes's entire poetics, present in poems about natural instincts and other signs of evolutionary processes. They are recognised both in the very large and in the very small, as suggested in these lines from 'Shackleton Hills', also discussed in Chapter 3:

> Stars sway the tree
> Whose roots
> Tighten on an atom.

This 'world tree' connects the smallest building blocks of nature, such as atoms, molecules and DNA, to the largest constellations in the universe, represented by the stars. For Hughes, these different scales of nature are interconnected and governed by the same universal and elemental powers. Humans form only a small part of this system, and are misguided when they think that they can control the forces of nature.

By depicting processes that are not readily apprehensible on a human scale, Hughes takes a biocentric rather than an anthropocentric standpoint. In many ways, the views expressed in his poems are similar to those of deep ecology. As Timothy Clark notes, for deep ecologists

> the essential problem is *anthropocentrism*, the almost all-pervading assumption that it is only in relation to humans that anything else has value. Deep ecologists urge a drastic change in human self-understanding: one should see oneself not as an atomistic individual engaged in the world as a resource for consumption and self-assertion, but as part of a greater living identity. All human actions should be guided by a sense of what is good for the biosphere as a whole. Such a *biocentrism* would affirm the intrinsic value of all natural life and displace the current preference of even the most trivial human demands over the needs of other species or integrity of place.
> (2011: 2)

Hughes promotes a similar understanding of the place of humans in relation to the non-human world.

Hughes's depictions of a universal, interdependent and connected ecosystem, expressed most explicitly in *River* and *Moortown Diary*, are, as noted by Jonathan Bate and also referred to in the introduction to this book, inspired by contemporary definitions of ecology. As Bate suggests, when the science of ecology gained traction and became increasingly popularised in the 1970s and 1980s, it offered a scientific framework for Hughes to which he could connect his ideas of ecological interdependence and environmental concern (though, as Bate also notes, natural scientists might 'not be entirely happy with Hughes's spiritual claims' (2000: 27) of a metaphysical connection between cells in the body, animals in the field and stars in the sky). For Hughes, the concept of ecology not only suggested such a connection of universal energies or natures but, furthermore, included the poet as especially positioned to 'tap into' and channel those energies.[2]

For Hughes, the connection to the natural environment is spiritual. As humans separate themselves from a close connection with the natural environment, they damage their own inner natures as well as their surroundings. This connection between inner and outer natures is a second expression of anti-anthropocentrism in Hughes's poetics, because it relies on the recognition of non-human natures within humans, necessary for a spiritual connection to the outer world. As discussed in Chapter 1, *Crow* enacts a rekindling of this spiritual connection to the natural world, which has been displaced by a Christian, technology-oriented worldview. In this light, *Crow* can be seen as part of a deep ecological movement, as far as this may be understood in the sense put forth by early ecocritic Andrew Brennan in 1988, in *Thinking about Nature*:

> deep ecologists argue that human flourishing or self-realisation requires a re-evaluation of our relationship with the rest of nature. This re-evaluation has taken various forms in the brief history of the movement, but a prevailing theme has been to urge the abandonment of our human-centred modes of thinking and valuing, and – more recently – to undertake a real identification with nature.
>
> (1988: 6)

Brennan's understanding of deep ecology relates it closely to Hughes's ecopoetics; he suggests for example that the views of deep ecologists often involve metaphysical claims, stating that 'Two common appeals are to idealism (the claim that the world is in some way mind-dependent) and to various kinds of global holism (the idea that all things are interdependent in a significant way)' (1988: 7). Hughes's notions of the poet as shaman, mediating a spiritual relationship between a human community and a world otherwise beyond human reach or comprehension, and of poetry and the imagination as ways of connecting to the goddess of nature (or the White Goddess described by Robert Graves in 1948) suggest a deep ecological connection to a fundamentally non-anthropocentric worldview. This strong version of deep ecology, characterised by a connection to the natural world that is at the same time metaphysical and non-transcendent, constitutes a special expression of what Bron Taylor describes as dark green religion (Taylor, 2010), as argued in Chapter 1. Relating to both deep ecology and dark green religion, Hughes's poetics could be characterised as a 'dark green poetry'.

Hughes's recognition of non-human nature within humans stresses similarities rather than differences between humans and other animals. The instincts and habits of different species that many of Hughes's animal poems describe are recognised as manifestations of slow evolutionary processes, making large-scale forces visible to the perspective of the single individual. Hughes's descriptions of predatory behaviour have caused some critics to accuse him of writing a 'poetry of violence'. However, as shown in Chapter 1, the violence that Hughes means to portray is not of a negative kind, according to his own explanation in the essay 'Poetry and Violence' (in Hughes, 1994). Rather it is a violence of a different order, inherent in all natural processes as a necessary part of the struggle for survival, and seen equally in destruction and creation. Performed in the way intended by their creator, these acts of violence are, in Hughes's poetics, 'natural' and amoral.

These natural forces are seen in some, but not all, human actions. Hughes recognises them in creative processes, such as the writing of poems. This link between evolution, animal instinct and human creativity is most directly illustrated in the poem 'Thrushes', where a small bird hunting for worms on an urban lawn is compared to a shark stabbing at a wound in its own side, and then to Mozart in the act of composing music. In 'Poetry and Violence', Hughes explains that these are different expressions of the same creative, elemental force, performed in perfect harmony with and according to the laws of nature. For Hughes, poetry can channel this powerful, potentially destructive force into a positive, regenerative energy. In so far as they capture this creative energy, Hughes's poems are themselves composed as expressions of a more-than-human nature.

The emphasis on similarities rather than differences between humans and other organisms suggests that Hughes's poetry can be read as posthumanist. Cary Wolfe states that the human/animal dichotomy depends on a definition of the human that 'is achieved by escaping or repressing [...] its animal origins in nature, the biological, and the evolutionary' (2010: p. xv). By emphasising precisely the ways in which humans are like other animals, and the fact that humans ultimately depend on a natural world that they cannot control, Hughes undermines the humanist human/animal dichotomy described by Wolfe. Furthermore, by suggesting ways of thinking and connecting to the natural world that have their origin in non-Western traditions and religions, Hughes distances himself from an Enlightenment thought tradition in particular, not just thematically but also through '*how* thinking confronts that thematics' (Wolfe, 2010: p. xvi).

A third expression of anti-anthropocentrism in Hughes's poetry is the recognition and portrayal of nature's own agency, apart and independent from humans. This is expressed in poems where the speaker attempts to move outside his own sphere of experience in order to see the world through different, non-anthropocentric eyes, and, in some sense, to speak 'for' or 'as' nature. In 'Salmon Eggs', for instance, the speaker tries to contemplate the river 'Till my eyes forget me'. A similar idea is approached in *Flowers and Insects*, as discussed in Chapter 5, in descriptions of agency in nature through recognition and representation of non-human semiosis and communication.

The anti-anthropocentric theme persists from Hughes's early poems to his later collections, but its mode of expression changes. While many of the early poems depict acts of violence or destruction, as in many of the animal poems, the later collections focus on processes of renewal, such as the annual cycles of the seasons in *River*, and of life and death in *Moortown Diary*. This thematic change from descriptions of animals in action towards observation of quieter, less dramatic and more continuous ecological relationships, from competition to symbiosis, is reflected by a corresponding change in form, from the forceful and assertive tone in the poems of *Crow*, for instance, towards the more reflexive and humble tone of *Moortown Diary* and *River*. This shift suggests a change of perspectives on the functions of ecology that reflects a corresponding change in the science of ecology, in line with Brennan's observation in *Thinking about Nature*, from the same decade as *River*, that 'far from competition being the ultimate determinant of life on earth, it seems that symbiotic associations are the *sine qua non* of any life support system' (1988: 129).

The development from poems about individual animals to depictions of larger ecosystems is also reflected in the format of Hughes's poems on the level of the collections, from his earliest collections where each poem formed a more or self-contained unit, to the later collections where the different poems form part of a larger story, as in *Crow*, *Gaudete*, *Remains of Elmet*, *Moortown Diary* and *River*. *Moortown Diary* stands out by including a role for humans in the ecosystem, thus adding a new dimension to the portrayal of ecological relations, perhaps suggesting a move away from the strongest forms of anti-anthropocentrism in Hughes's poetics.

Heaney – language, ecology and the question of the (non-)native

As discussed in Chapter 4, many of Heaney's poems revolve around how linguistic environments reflect and co-evolve with their natural settings. Poems like 'A Backward Look' and 'A Herbal' suggest how a particular name for a species intertwines with the species itself. Reflecting a creative imagination shaped by discourses of anti-colonial thought, Heaney relates both the linguistic and the natural environment to questions of nativeness, recognising strangers that intrude on, challenge and compete with native perspectives and species, sometimes threatening existing species but sometimes also creating possibilities for new ecosystems where old and new species can coexist.

The intertwining of the linguistic and the natural world animates the poem 'Canopy' (Heaney, 2010: 44–5), from *Human Chain*. In this poem trees in the Harvard yard, through which the speaker is walking, resonate with voices from 'everywhere'. This is not in the speaker's imagination, but the result of hidden amplifiers in the trees' canopies. The poem suggests not only that the local, rooted trees are connected to whispers from afar, but also that nature resonates, in this case literally, with human languages, cultures and ideas. It also illustrates how human societies and cultures are embedded in environments of speech and languages that form 'ecologies' of their own. Moreover, the poem illustrates how human voices speak out of a natural setting:

> It was the month of May,
> Trees in Harvard Yard
> Were turning a young green.
> There was whispering everywhere.
>
> David Ward had installed
> Voice-boxes in the branches,
> Speakers wrapped in sacking
> Looking like old wasps' nests
> [...]
> It was like a recording
> Of antiphonal responses
> In the congregation of leaves.

'Canopy' indicates how languages cannot merely be described as linguistic environments, but actually take part in the ecology in the sense noted by Lawrence Buell. 'Genres and texts are themselves arguably "ecosystems", not only in the narrow sense of the text as a discursive "environment", but also in the broader sense that texts "help reproduce sociohistorical environments" in stylized form' (Buell, 2005: 44). The same notion is expressed as a process of 'environing' by Paul Warde and Sverker Sörlin in the introduction to *Nature's End*, describing how people turn 'nature', something quite distinct from a human or cultural realm, into 'environment', nature as a social category, co-produced and historicised by humans (Sörlin and Warde, 2009: 8). In 'Canopy', and elsewhere in Heaney's poetry, the human and natural worlds are always portrayed in mutual processes of 'environing'.

In Heaney's linguistic ecology, there are native as well as alien words, and some that are threatened by extinction. Heaney's poetics also include frequent examples of words that exhibit particular relationships with place, through their origins and dialectal variations. As discussed in Chapter 4, the linguistic environment, like the natural one, is for Heaney an environment to cherish, where iteration and reuse of individual words and expressions, as in the last part of 'An Old Refrain', are attempts to do just that: words become 'gleanings and leavings' to be kept 'in the combs / of a fieldworker's archive', as expressed in the earlier poem 'The Backward Look'.

The presence of non-natives in the poems is complicated by the fact that they are written in English, the alien tongue. Irish words in an otherwise English poem take on a double standard, simultaneously native and alien, while the English words are intruders that are at the same time already present 'within' the native ecology. The poem becomes a hybrid, a place for preservation and reintroduction of species in an environment that is at the same time native and foreign, where the poem itself becomes an environment in which multiple languages, traditions and cultures coexist.

Heaney's focus on native places, languages and dialects to some extent represents what Clark describes as a 'bioregional project of "reinhabitation"' (2011: 125). However, the presence of non-natives in the form of English and other foreign references are signs of a countervailing internationalism, a current that generally runs stronger than bioregionalism in Heaney's poetics, especially in the later collections. The combination of and tensions between bioregional and international trends suggest images that combine local with global perspectives, relating to the kind of environmental cosmopolitanism that Heise describes and that are discussed in Chapter 6.

The presence of strangers in the (natural as well as linguistic) environment also counteracts nativist and bioregional emphases by introducing non-essentialist views of cultures and species that relate to queer ecocritical perspectives. As suggested in Chapter 4, the elaboration on species and species' names in 'An Old Refrain' suggests a willingness to include strange and foreign words in order to form a hybrid of English and Irish linguistic traditions. That both English and Irish words are strangers in some respect suggests that they can be described as the kind of 'strange strangers' that Timothy Morton refers to in his essay about 'Queer Ecology': 'Their familiarity is strange, their strangeness familiar' (2010: 277).

Another layer of strangeness versus familiarity can be seen in 'An Old Refrain' if the linguistic and the natural are considered together, so that a familiar plant becomes strange when it is named by a foreign language, perhaps especially if that language belongs to a coloniser. The strangeness introduced is then similar (though uninvited) to the strangeness experienced in 'Making Strange', also discussed in Chapter 6, where Heaney brings a friend from abroad to Ireland, and is faced by this person standing next to his own father, who suddenly seems strange to the speaker through the foreigner's eyes. The presence of these multiple layers of strangeness in Heaney's poetics, and the instability of the concept of the native, reinforces the trend away from nativism and bioregionalism and towards internationalism and cosmopolitanism that is recognised also on other grounds in Heaney's poetics in Chapter 6.

Hughes, Heaney and the different natures of ecopoetics

The concept of an ecological poetics suggests that there is a special relationship between ecology or environmentalism and the formal qualities of poetry; that ecopoetry is not merely 'about' nature in a thematic sense, but actually addresses or reflects ecological relationships in a different, intrinsic way. Angus Fletcher argues for a particularly strong form of ecopoetics in *A New Theory for American Poetry*, stating that 'poetry takes environmentalist concerns to a higher level':

> Unlike most prose discourse, poetry expresses close personal involvements, and hence pertains to the way we humans respond, on our own, to environmental matters. [...] An art like poetry that enhances the presence of the individual is bound to be central in showing how we should understand our environmental rights and obligations. The issue then is this, what is my own response to my surrounding?
>
> (2004: 3–4)

For Fletcher, poetry is not merely a reflection or representation of nature, it actually *is* nature; the poem constitutes an 'environmental form' (Fletcher, 2004: 6). It is an expression of the same forces that create the physical world, and is part of the same global ecosystem to which all other forms of life ultimately belong.

Developing this argument, Fletcher identifies what he calls 'the *environment-poem*, a genre where the poet neither writes *about* the surrounding world, thematizing it, nor analytically represents the world, but actually shapes the poem to *be* an Emersonian or esemplastic circle'. Invented by Walt Whitman, environment-poems, according to Fletcher, 'aspire to surround the reader so that to read them is to have an experience much like suddenly recognizing that one actually has an environment, instead of not perceiving the surround at all'; Whitman's rhapsody, for example, Fletcher argues, 'perpetually liminal, acquires a Thoreauvian wildness' (2004: 9). Though Fletcher limits his argument to poetry,

Buell suggests that it could be expanded to include other genres as well, and that such environment-texts could help recognise 'how the social landscape figures as part of total landscape' (2005: 51).

The relationship between ecology and poetry is also discussed by Hubert Zapf in the essay 'Literary Ecology and the Ethics of Texts', and by John Felstiner in *Can Poetry Save the Earth?* Taking a different approach than Fletcher, Zapf is interested in the relationship between biological and cultural ecologies, which he characterises as a relationship of 'interdependence-yet-difference' (2008: 851). He suggests that cultural ecologies that involve 'the internal landscapes produced by modern culture and consciousness are equally important for human beings as their external environments' (2008: 852). In so far as they address how the experiences that shape those landscapes are expressed in language and images that rely on sensory relations and responses to the natural environment, cultural ecologies are intertwined with biological ecologies. According to Zapf, this interrelationship between language, imagination and nature suggests that literature is an especially compelling form of cultural ecology, because it can express and explore the changing relations, the 'nonlinear complex feedback relationships', between nature and culture. Literature that explores the relationship between nature and culture in this sense Zapf terms 'literary ecologies'.

Like Hughes, Felstiner emphasises poetry's ability to 'quicken awareness' of our natural surroundings as a first step towards developing the will to 'lighten our footprint in a world where all of nature matters vitally', according to the sentiment 'First consciousness then conscience' (2009: p. xiii). Felstiner's statements that 'poetry more than any other kind of speech reveals the vital signs and warning signs of our tenancy on earth', and that 'Poems make us stop, look, listen long enough for imagination to act, connecting, committing ourselves to the only world we've got' (Felstiner, 2009: 4, 11) are similar to Hughes's comparison in 'Poetry in the Making' between a poet and a fisherman, staring at his float. What these two activities have in common, Hughes suggests, is the ability to focus one's attention on a single thing or thought, and hold it there for so long that all distractions dissolve and one can 'enter one of the orders of bliss'. In this state of mind, Hughes argues, the poet/fisherman can become aware of the world beyond the image/float:

> You are aware, in a horizonless and slightly mesmerized way, like listening to the double bass in orchestral music, of the fish below in there in the dark. At every moment your imagination is alarming itself with the size of the thing slowly leaving the weeds and approaching your bait. Or with the world of beauties down there, suspended in total ignorance of you. And the whole purpose of this concentrated excitement, in this arena of apprehension and unforeseeable events, is to bring up some lovely solid thing, like living metal, from a world where nothing exists but those inevitable facts which raise life out of nothing and return it to nothing.
>
> (Hughes, 1994: 19)

Those 'inevitable facts' signify something similar to the 'only world we've got' referred to by Felstiner. The process that Hughes describes of training the imagination to 'settle' on that world long enough to form a 'deeper' connection to it than we experience in our everyday lives results in the kind of ecopoem that Felstiner describes. As Garrard suggests in a review of Felstiner, this notion of ecopoetics represents 'a type of ecocritical Russian formalism that promotes the moral value of poetry's *ostranenie* (defamiliarisation)' (2010: 14).

Like Fletcher, Felstiner stresses the importance of the individual as the basis for poetry's relevance to environmentalism, stating that 'The essential choices, ticklish for government and industry, fall to us first as individuals in our eating, housing, clothing, childbearing, transport, recreation, voting' (2009: 13). The emphasis on the importance of the individual's role in the relationship between poetry and the environment is contradicted by Clark's claim that environmental thinking is so difficult precisely because its scope is insufficient as long as it starts with the individual – that 'Scale effects are straightforward to exemplify but impossible to apprehend in any particular individual case'. Clark points out that climate change in particular is not about the individual, but rather 'a matter of *context*', and that this is difficult for ecocritics (and others) to deal with, other than in abstract terms:

> much ecocriticism takes the individual attitude as its starting point and then argues for a change in the choices which that individual makes. Thus, it is hoped, the growth of an 'ecological awareness' through the study of environmentalist non-fiction, eco-poetry or real ventures into the wild, will be somehow sufficient to produce an ecologically viable society. Such thinking effectively recognises that climate change enacts a drastic reconfiguration of given distinctions of public and private but, without more sustained work on the nature of the state, ideology, modes of production etc, still seeks to engage it solely in terms of individual attitude and choice. The focus on the individual, whether as green consumer, a reader of an ecocritical argument, or as a backpacker, reinforces the illusion that reality and power remain a matter of individuals pursuing their rights and opinions.
>
> (2010: 135)

Climate change, according to Clark, means that we are 'suddenly in need of confronting how current modes of thinking and acting are inadequate or anachronistic' (2010: 141). In its most serious implications, environmentalism, including literature and criticism, is about how the human population responds to environmental problems, more than about how each of us responds on our own.

Fletcher's affirmation of the importance of the individual, and his statement that 'the strong message of [the environment-poem], if it needed one, would be that a good society must become a self-organizing system, without too much top-down control', is subject to the risk described by Clark of turning environmental politics into a question of the moral choice and responsibility of the individual, thereby reinforcing 'a culture of narcissistic individualism already implicated in consumer democracy and environmental danger' (2010: 144–5). Clark argues that ecocritics need to engage with questions involving environmental racism and elitism and health as matters of social justice on a level that is not centred around the individual.

Clark's definition of ecopoetry is different from both Fletcher's and Felstiner's, though it could probably include the latter's description. It focuses on the poem's ability to challenge established modes of thought and offer new types of images for relating to environmental problems of non-individual proportions and scales:

> a loosely 'ecological' poetic emerges in the development and extension of modernist techniques that had been initially pioneered in the first four decades of the twentieth century. At issue is an aesthetic interested in formal experimentation and the conception of the poet or poem as forming a kind of intellectual or spiritual frontier, newly coupled with a sense of the vulnerability and otherness of the natural world, distrust of a society dominated by materialism and instrumental reason, and sometimes giving a counteraffirmation of non-western modes of perception, thought or rhetorical practice. The poem is often conceived as a space of subjective redefinition and rediscovery through encounters with the non-human.
>
> (Clark, 2011: 139)

In different ways, both Hughes and Heaney can be described as ecopoets in light of Felstiner's and Clark's definitions. Both poets' works 'live on the sensory shock of things', and depend on 'seeing the things of our world afresh by saying them anew' (Felstiner, 2009: 2, 3). Hughes, like Felstiner, stresses the importance of a heightened awareness of non-human nature and the role of the imagination, trained by reading and writing poetry, for connecting to the natural world. Hughes's animal poems, in particular, which question and challenge definitions of human nature, enact encounters with non-human nature. Intertwining linguistic with natural environments, Heaney's poems illustrate the kind of literary ecology characterised by Felstiner as exploring the interdependencies between cultural and biological ecologies.

Hughes's ecopoetics, which strives towards speaking for, or even as, nature and 'channelling' the energies of the natural world, reflects Fletcher's claim that poetry 'gives voice to the unbreakable link between nature and humanity, since poetry, our imaginative making, seems to participate in nature' (2004: 4). In Hughes's case, this 'unbreakable bond' does not suggest a bond between nature and humans that is different from other ecological relationships, but rather indicates the presence of nature within humans, with poetry being one of its potential outlets or expressions. References in Hughes's poems to very large- as well as very small-scale processes, and to natural processes within and without the individual, relate human experiences to an interconnected, global ecosystem, in which the poems themselves participate.

Heaney's ecopoetics is of a different kind than Hughes's. Rather than moving away from anthropocentric perspectives, Heaney suggests an 'ecology' that incorporates natural as well as cultural elements, especially by combining and intertwining natural with linguistic environments. Heaney's poems not only record or reflect such processes, but in themselves contribute to this natural/linguistic environment in ways similar to those outlined by Zapf with regard to biological and cultural ecologies, including linguistic reflection of sensory experiences of the natural environment and interdependencies between natural and political/cultural processes, as discussed in Chapter 4.

The tension between bioregionalism and internationalism in Heaney's poetics involves a sometimes fraught relationship between local loyalty and cosmopolitan freedom and self-assertion. In addition to the examples discussed in Chapter 6, poems in *Station Island* also reflect this tension. In 'Station Island xii' (Heaney, 1984: 92–4), the last of the 'Station Island' poems, Michael Cavanagh notes that the speaker encounters 'the ghost of James Joyce', who tells him, with a 'voice eddying with the vowels of all rivers', to set off on his own, and trust his own voice (2009: 38):

> 'Your obligation
> is not discharged by any common rite.
> What you must do must be done on your own
>
> so get back in harness.
> [...]
> Take off from here. And don't be so earnest,
>
> let others wear the sackcloth and the ashes.
> Let go, let fly, forget.
> You've listened long enough. Now strike your note.'

The poem describes not only the speaker's difficulty in trusting his own voice in relation the earlier writer, but also the letting go of a sense of responsibility of writing 'for' Northern Ireland. The poem claims the speaker's right to 'fill the element / with signatures of [his] own frequency', a necessity for the poetic imagination: 'You lose more of yourself than you redeem / doing the decent thing'. The poem urges the speaker to accept and claim that

> The English language
> belongs to us. You are raking at dead fires,
>
> a waste of time for somebody your age.

On grounds of poetic as well as personal freedom, the poem justifies moving out of the local, Irish context, but it also contains an element of self-criticism for not being able to do so. Abandoning the familiar is associated with both fear and freedom:

It was as if I had stepped free into space
alone with nothing that I had not known
already.

A similar claim for artistic freedom is further outlined in 'On the Road' (Heaney, 1984: 119–21), the last poem of *Station Island*, where the speaker is

up and away

like a human soul
that plumes from the mouth.

The poem describes the necessity of being in motion, of finding new places and perspectives by following different paths. At the end of his journey, the speaker imagines, he will find a hidden place, far away from his origins:

I would migrate
through a high cave mouth
[...]
to the deepest chamber.

In this magical place, he imagines he will find inspiration:

I would meditate
that stone-faced vigil

until the long dumbfounded
spirit broke cover
to raise a dust
in the font of exhaustion.

The emphasis on motion and on seeing things from the outside in order to both experience new things and gain a different understanding of the familiar captures an important dimension of developing an environmental world citizenship (as discussed by Heise (2008) and in Chapter 6), suggesting that local nature cannot be comprehended only from within, from a bioregional perspective, as suggested by Fletcher, but also needs the outsider's view.

The relationship in 'On the Road' between immediate, physical nature and the outsider's view suggests a widening ecological thought that can begin to include, as Clark suggests, components and scales that are not immediately accessible to the single individual. The trend towards an increasing international and global focus in Heaney's later poems compared to his early collections is, as discussed, perhaps better described less as a trend towards internationalism as a replacement for local or regional perspectives than as a trend towards images that are capable of relating simultaneously to notions of varying scale, from the local and regional to the national, global and universal. Like Hughes's poetry, Heaney's poems also address questions of scale in this sense.

The problem of scale is most acute in issues relating to climate change, which, as Clark points out, presents a challenge to ecocritics of 'keeping pace intellectually with an event whose scale, complexity and incalculability is such to resist representation or being conceptualised' (2010: 132). Clark suggests that the challenges that climate change pose to the ability of humans to conceptualise issues of such scale and uncertainty render attempts to meet environmental crises with ideas of better management insufficient, as the idea of management itself assumes that humans can understand and control the natural world, something which an issue as complex and unpredictable as climate change indicates is impossible. He argues that, in light of our new understanding of 'the finitude of the earth', we need to trace and rethink the anthropocentric framing of that understanding, to include other dimensions of the natural world than those that can be described accurately and finitely through 'calculable technic' (2010: 133–4). This limited view suggests a sense of closure that does not take into account any agency other than that of humans:

> The condition of closure renders anachronistic inherited economic, political practices and modes of judgement without acceptable alternatives appearing in their place. The epoch whose intellectual closure is now visible, the 'flat earth' epoch so to speak, inaugurates the need to think a bounded space in which the consequences of actions may mutate to come back unexpectedly from the other side of the planet. The 'environment' is no longer thinkable as an object of 'crisis' for us to decide on or manage: it ceases being only a passive ground, context and resource for human society and becomes an imponderable agency that must somehow be taken into account, even if we are unsure how. (2010: 134)

Hughes suggests that the creative process of reading and writing poetry is a way of realising and considering the agency of the natural world. By connecting to rather than controlling forces of nature, humans would, in Hughes's view, gain a better, more humble understanding of their place in the global ecosystem. Hughes's poetics thus suggests not merely a modification or adaptation of current environmental values and thoughts, but a drastic change to a new, spiritual and ecocentric view of nature–human relations. Clark describes the 'intellectual closure' of modernity (defined 'as the assumption that the natural world exists for human ends, the dominance of liberal-democratic systems of government embedded in market capitalism, and the privilege given to scientific knowledge as the solely reliable guide in managing the social and natural worlds') as a cul-de-sac, brought into effect by the scope, scale and complexity of climate change (2010: 142). The description is similar to Hughes's suggestion of 'the dead end of a wrong direction' from 'Top Withens', referring to the end of the relationship between exploring, exploiting humans and a passive, infinite nature as it has developed at least since the last Ice Age.

The notion of nature as a passive, stable entity is also addressed by Morton, who states in *Ecology without Nature* that this 'idea of "nature" which so many hold dear will have to wither away in an "ecological" state of human society'. Morton wants to replace the concept of nature with a particular notion of ecology:

> 'Ecology without nature' could mean 'ecology without a *concept* of the natural'. Thinking, when it becomes ideological, tends to fixate on concepts rather than doing what is 'natural' to thought, namely, dissolving whatever has taken form. Ecological thinking that was not fixated, that did not stop at a particular concretization of its object, would thus be 'without nature'.
> (2007: 1)

Morton also suggests that 'it is in art that the fantasies we have about nature take shape – and dissolve' (2007: 24). Brennan anticipates Morton's argument already in his study from 1988, noting the dangers of equating the concept of 'natural' behaviour with innate qualities of the organism and stating that, given the confusion surrounding the term 'natural', 'a case could be made for dropping the term from our descriptions of phenomena' (1988: 88). Brennan points out that, though there are some connections between 'natural' behaviour and inherited qualities, such as physical characteristics making certain diets more 'natural' than others, attempts to translate this connection into moral guidelines quickly become problematic. Extended beyond strictly biological traits, Brennan states, such reasoning can be used to argue that 'everything that happens is natural' (1988: 90).

Neither Hughes nor Heaney make a clear distinction between nature and ecology as differentiated by Morton, but the distinction between nature as something fixated and 'out there' and ecology as something less constant and easier to adapt to new ideas of ecosystems and biomes that are integrated with human societies can be used to point to an important difference between Hughes's and Heaney's ecopoetics.

For Hughes, though nature has agency, it remains a stable point of reference, while human imagination and conceptualisation falter both in judgement and in the ability to perceive, comprehend and live according to natural processes and laws. Hughes's poems strive to portray that which lies beyond social and cultural constructions of nature, through anti-anthropocentric descriptions and points of view. Morton suggests that nature can refer either to a substance or an essence (and notes that this is 'One of the basic problems with nature', 2007: 16). For Hughes, nature is primarily an essence, and it is this essence that the poems search for and aim to describe, and identify in the DNA of humans, in the behaviour of other animals, in all ecological relationships and in the natural processes that shape the world beyond human influence. In Hughes's poetics, different substances of nature are, as Morton puts it, 'just a variation in their atomic structure' (2007: 18). For Hughes, this essence is sacred without being transcendental or anthropocentric.

Hughes's ecopoetics contradicts Morton's statement that environmental literatures 'encapsulate a utopian image of nature which does not really exist – we have destroyed it' (2007: 24). The essence that Hughes describes can be pushed back, as in 'Crow's Undersong' (discussed in Chapter 1), but it cannot be destroyed, only its different materialisations can be, as seen in poems where Crow is obliterated only to be immediately resurrected (see e.g. 'Magical Dangers' or 'Glimpse'). As an evolutionary trickster, Crow embodies a different nature concept than the utopian image that Morton refers to. Morton notes that 'for many cultures nature is a trickster' (2007: 31), and this is certainly the case for Hughes. This means that for Hughes Crow embodies not just evolution but the concept of nature in a wider sense. Rather than thematising nature, *Crow* is an enactment of the nature concept in Hughes's poetics.

Heaney does not, like Hughes, search for a nature that is untainted by humans. Instead, he is interested in the mutually influencing relationship between nature and culture and in the details of how nature is constructed by human language, values, culture and history. While Hughes describes 'simply nature', Heaney portrays the human and the non-human world as intertwined. In *Song of the Earth*, Jonathan Bate states that, like Wordsworth, Heaney

> sees into the life of things and he yearns for the rural places in which he grew up. But, like Yeats, he has been hurt into poetry by 'mad Ireland'. In *Seeing Things* (1991), his finest volume of mature work, he revisits Glanmore, the farm of his childhood, but he never fully gets back to nature. He squares up to the Januslike quality of the poet – singer of earth, exile from earth – remains warily on guard as he crisscrosses between culture and nature.
> (Bate, 2000: 203)

Perhaps he does not even crisscross between nature and culture, sees them as inseparable. Commenting further on the lines from Heaney's poem 'Squarings': 'I took a turn and met the fox stock-still, / Face-to-face in the middle of the road', from *Seeing Things*, Bate compares Heaney to Hughes:

> at the heart of the poem the speaker is entered by 'wildness' but cannot return to the wild because he is contained within his car. Furthermore, the fox is not Heaney's habitual animal. It belongs to his friend, and a key influence on his first volume, *Death of a Naturalist* (1966): Ted Hughes. Hughes was a truly feral poet, scavenging like a fox on the margins of urban modernity. His poetry has the hot stink of animal flesh, whereas this fox of Heaney's is cunningly hedged with literariness [...] His compulsion to cross back through the fox's startled eye is a symptom of his loss of unity with nature.
> (2000: 203)

A corresponding difference is seen in Heaney's increasingly cosmopolitan outlook, compared to Hughes's insistence on immediate experiences of his surroundings, also noted by Bate:

> In contrast to Seamus Heaney, who began as a poet of turf, bog and locality, but has become ever more cosmopolitan in his interests, Hughes dug himself

> in a smallholding in the far west of England. In his later poetry, he is always driving away from London, back to the farm. He rarely went abroad, and when he did the places which inspired him were the Australian outback and the wilderness of Alaska.
>
> (2000: 29)

These physical journeys reflect Hughes's and Heaney's poetic journeys and the development of their respective ecological thinking.

Clark points out that the questions that face ecocritics in the 2010s are of an entirely different, or at least additional, order to those that informed early ecocriticism in the 1990s. Complex climate change issues, requiring new and challenging theoretical and conceptual approaches, have added to concerns regarding preservation of individual species or local pollution. As a result, ecocriticism is rapidly changing and diversifying. For ecocritics, Clark notes, 'there seems no off-the shelf oppositional stance ready to be used or adapted, only a great deal of new work to be done' (2010: 147).

Among the new critical and theoretical directions initiated, a general trend can be recognised away from what Kate Soper has called a nature-endorsing view to a 'nature'-sceptical view (Soper, 1995), from a commitment to mimesis to engagement with ideas about the complex relationships between different natures and different human societies. This change has profound effects on the notion of what ecopoetics can or should be, leaning away from poetry that thematises nature towards poems that offer, to use Heaney's formulation from a different context, 'images and symbols adequate to our predicament' (Heaney, 1980: 57–8).

Notes

1 A version of this chapter, co-authored with Greg Garrard, is published as '"Images Adequate to our Predicament": Ecology, Environment and Ecopoetics', *Environmental Humanities*, 5: 35–53(2014).

2 Hughes explains this notion in, for instance, 'Ted Hughes and Crow: An Interview with Egbert Faas', in Faas, 1980.

References

Bate, J. (2000) *The Song of the Earth*, London: Picador.

Brennan, A. (1988) *Thinking about Nature: An Investigation of Nature, Value and Ecology*, London: Routledge.

Buell, L. (2005) *The Future of Environmental Criticism: Environmental Crisis and Literary Imagination*, Malden, MA: Blackwell Publishing.

Cavanagh, M. (2009) *Professing Poetry: Seamus Heaney's Poetics*, Washington, DC: Catholic University of America Press.

Clark, T. (2010) 'Some Climate Change Ironies: Deconstruction, Environmental Politics and the Closure of Ecocriticism', *Oxford Literary Review*, 32/1: 131–49.

Clark, T. (2011) *The Cambridge Introduction to Literature and the Environment*, New York: Cambridge University Press.

Felstiner, J. (2009) *Can Poetry Save the Earth: A Field Guide to Nature Poems*, New Haven, CT: Yale University.
Fletcher, A. (2004) *A New Theory for American Poetry: Democracy, the Environment, and the Future of Imagination*, Cambridge, MA, and London: Harvard University Press.
Garrard, G. (2010) 'Ecocriticism', *The Year's Work in Critical and Cultural Theory*, 18/1: 1–35.
Heaney, S. (1972) *Wintering Out*, London: Faber & Faber.
Heaney, S. (1980) *Preoccupations: Selected Prose, 1968–1978*, New York: Farrar, Straus & Giroux (first published by Faber & Faber).
Heaney, S. (1984) *Station Island*, London: Faber & Faber.
Heaney, S. (1996) *Seeing Things*, London: Faber & Faber.
Heaney, S. (2010) *Human Chain*, London: Faber & Faber.
Heise, U. (2008) *Sense of Place and Sense of Planet: The Environmental Imagination of the Global*, New York: Oxford University Press.
Hughes, T. (1994) *Winter Pollen: Occasional Prose*, London: Faber & Faber.
Hughes, T. (2003) *Collected Poems*, London: Faber & Faber.
Morton, T. (2007) *Ecology without Nature: Rethinking Environmental Aesthetics*, Cambridge, MA: Harvard University Press.
Morton, T. (2010) 'Queer Ecology', *PMLA*, 125/2: 273–82.
Soper, K. (1995) *What is Nature? Culture, Politics and the Non-Human*, Oxford: Blackwell.
Sörlin, S., and Warde, P. (2009) 'Making the Environment Historical – An Introduction', in S. Sörlin and P. Warde (eds), *Nature's End: History and the Environment*, pp. 1–22, Basingstoke: Palgrave Macmillan.
Taylor, B. (2010) *Dark Green Religion: Nature Spirituality and the Planetary Future*, Berkeley, CA: University of California Press.
Wolfe, C. (2010) *What is Posthumanism?* Minneapolis, MN: University of Minnesota Press.
Zapf, H. (2008) 'Literary Ecology and the Ethics of Texts', *New Literary History*, 39/4: 847–68.

Index

www.ingramcontent.com/pod-product-compliance
Ingram Content Group UK Ltd.
Pitfield, Milton Keynes, MK11 3LW, UK
UKHW011017180926
475701UK00004B/796
* 9 7 8 1 1 3 8 7 4 3 5 8 8 *